Lecture Notes in Mathematics
A collection of informal reports and seminars
Edited by A. Dold, Heidelberg and B. Eckmann, Zürich

41

Robin Hartshorne
Harvard University, Cambridge, Mass.

Local Cohomology

A seminar given by A. Grothendieck
Harvard University
Fall, 1961

1967

Springer-Verlag · Berlin · Heidelberg · New York

Library of Congress Catalog Card Number 67-29865. Printed in Germany. Title No. 7361.

CONTENTS

Introduction

What follows is a set of lecture notes for a seminar given by A. Grothendieck at Harvard University in the fall of 1961. The subject matter is his theory of local (or relative) cohomology groups of sheaves on preschemes. This material has since appeared in expanded and generalized form in his Paris seminar of 1962 [16] and my duality seminar at Harvard in 1963/64 [17]. Furthermore, it may appear in the later sections of his "Elements," chapter III [5]. However, I have thought it worthwhile to make these notes available again, since a short, elementary treatment of a subject is often the best introduction to it. The text is essentially the same as the first edition, except for minor corrections and an expanded bibliography.

The study of local cohomology groups has its origin in the observation, already implicit in Serre's paper FAC [10], that many statements about projective varieties can be reformulated in terms of graded rings, or complete local rings. This allows one to conjecture and then prove statements about local rings, which then may be of use in obtaining better global results. Thus the finiteness theorems of Serre for coherent sheaves on projective varieties become statements that certain local cohomology modules are "cofinite." Similarly the duality theorem for projective varieties becomes a duality theorem for local cohomology modules. This approach was developed further by Grothendieck in his 1962 seminar [16], where he studies local and global Lefschetz theorems, relating π_1 and Pic of a variety to π_1 and Pic of its hyperplane sections.

In Section 1 we define the local cohomology groups of an abelian sheaf F on a topological space X, with respect to a locally closed subset Y, as the right derived functors of the functor $\Gamma_Y(F)$, the sections of F "with support in Y."

In Section 2 we give the first consequences of this theory when applied to preschemes: The main result is Theorem 2.8, which interprets the local cohomology groups as a direct limit of Ext's.

In Section 3 we give a self-contained exposition

of the notion of depth (or homological codimension) and relate it to the local cohomology groups (Theorem 3.8).

Section 4 contains some "general nonsense" on functors defined on the category of modules over a Noetherian ring. In particular, we discuss dualizing functors and dualizing modules.

In Section 5 we give some miscellaneous applications and some technical results useful in the next section. (In the lectures, most of this material came before that of Section 4, and it is perhaps better to read it first, since it is closely related to Section 3).

Section 6 contains the duality theorems which are central results of the local cohomology theory.

R. Hartshorne
Cambridge, Mass.
July, 1967

§ 1. Definition and Elementary Properties of the Local Cohomology Groups

For preliminaries of homological algebra we refer to Godement [4] or Grothendieck [6]. The results of this section will be valid for arbitrary topological spaces and arbitrary sheaves of abelian groups over them.

Let X be a topological space, let Z be a locally closed subspace of X, and let F be an abelian sheaf on X. (We recall that a subspace Z of a topological space X is <u>locally closed</u> if it is the intersection of an open and a closed subset. By an <u>abelian sheaf</u> on a topological space X we mean simply a sheaf of abelian groups on X. The category of all abelian sheaves on X will be denoted by $\mathcal{C}(X)$.)

Choose an open subset $V \subseteq X$ such that $Z \subseteq V$, and Z is closed in V. This is possible since Z is locally closed in X. Let $\Gamma_Z(X, F)$ be the subgroup of $F(V)$ consisting of all those sections of F whose support is contained in Z. One checks immediately that $\Gamma_Z(X, F)$ is independent of the subset V chosen above. Moreover, one sees also that the functor

$$F \rightsquigarrow \Gamma_Z(X, F)$$

from $\mathcal{C}(X)$ to (Ab) is left-exact. We call $\Gamma_Z(X, F)$ the <u>sections of F with support in</u> Z. Note that if Z is closed, it is precisely this; if Z is open in X, $\Gamma_Z(X,F) = F(Z)$.

Let X, Z, F, V be as above. Then if U is any open subset of X, the natural restriction homomorphism

$$F(V) \to F(V \cap U)$$

induces a homomorphism

$$\Gamma_Z(X,F) \to \Gamma_{Z\cap U}(U, F|U) \quad .$$

Thus we may consider the presheaf

$$U \rightsquigarrow \Gamma_{Z\cap U}(U, F|U) \quad .$$

One verifies immediately that in fact this presheaf is a sheaf, and we will denote it by $\underline{\Gamma}_Z(F)$. As with Γ_Z, one finds that the functor

$$F \rightsquigarrow \underline{\Gamma}_Z(F)$$

is left-exact from $\mathcal{C}(X)$ to $\mathcal{C}(X)$.

<u>Definition</u>. Let X be a topological space, Z a locally closed subspace, and F an abelian sheaf on X. Then the right derived functors of Γ_Z and $\underline{\Gamma}_Z$, respectively, are denoted by $H^p_Z(X,F)$ and $\underline{H}^p_Z(F)$, $p = 0, 1, \ldots$, respectively, and are called the <u>cohomology groups</u> (resp., <u>cohomology sheaves</u>) <u>of</u> X <u>with coefficients in</u> F <u>and supports in</u> Z.

<u>Remark</u>: Using the notion of a sheaf extended by zero outside of a locally closed subset [4, II 2.9], we can give a different

interpretation of $\Gamma_Z(F)$ and $\underline{\Gamma_Z}(F)$. Let $\mathbb{Z}_Z$ be the constant sheaf of integers on Z, and let $\mathbb{Z}_{Z,X}$ be this same sheaf extended by zero outside of Z. Then

$$\Gamma_Z(X, F) = \mathrm{Hom}\,(\mathbb{Z}_{Z,X}, F)$$

and

$$\underline{\Gamma_Z}(F) = \underline{\mathrm{Hom}}\,(\mathbb{Z}_{Z,X}, F) \quad .$$

where $\underline{\mathrm{Hom}}$ denotes the sheaf of germs of homomorphisms.

The rest of this section will be devoted to various properties of these cohomology groups and sheaves with supports in a locally closed subspace.

Proposition 1.1. Let Z be locally closed in X. Then

a) For any $F \in \mathfrak{S}(X)$,

$$\Gamma_Z(X, F) = H^o_Z(X, F)$$

$$\underline{\Gamma_Z}(F) = \underline{H^o_Z}(F) \quad .$$

b) If $0 \to F' \to F \to F'' \to 0$ is an exact sequence of abelian sheaves on X, then there are long exact sequences

$$0 \to H^o_Z(X, F') \to H^o_Z(X, F) \to H^o_Z(X, F'') \xrightarrow{\partial}$$

$$H^1_Z(X, F') \to H^1_Z(X, F) \to \ldots$$

and

$$0 \to \underline{H}^o_Z(F') \to \underline{H}^o_Z(F) \to \underline{H}^o_Z(F'') \xrightarrow{\partial} \underline{H}^1_Z(F') \to \underline{H}^1_Z(F) \to \ldots,$$

where the connecting homomorphisms ∂ behave functorially.

c) If $F \varepsilon \mathfrak{C}(X)$ is an injective object of the category, then $H^p_Z(X, F) = 0$ for $p > 0$, and $\underline{H}^p_Z(F) = 0$ for $p > 0$.

Proof. These properties follow from the fact that H^p_Z and $\underline{H}^p_Z$ are derived functors of Γ_Z and $\underline{\Gamma}_Z$. In fact, a, b, c characterize the derived functors.

Proposition 1.2. With Z, X, F as above, for each $p \geq 0$, $\underline{H}^p_Z(F)$ is the sheaf associated to the presheaf

$$U \rightsquigarrow H^p_{Z \cap U}(U, F|U) \quad .$$

Proof. This follows formally from the definitions. Set $\underline{H}^{p'}(F)$ equal to the sheaf associated to the presheaf above. Then, since the operation of taking associated sheaves is an exact functor, the $\underline{H}^{p'}(F)$ form a connected sequence of functors. Moreover, for $p = 0$, $\underline{H}^{o'}(F)$ = the sheaf associated to the presheaf $(U \rightsquigarrow \Gamma_{Z \cap U}(U, F|U))$, $= \underline{\Gamma}_Z(F)$. If F is injective, so is $F|U$ for any U; hence, all the $H^p_{Z \cap U}(U, F|U) = 0$ for $p > 0$. Therefore, also, $\underline{H}^{p'}(F) = 0$ for $p > 0$. But these properties characterize the right derived functors of $\underline{\Gamma}_Z(F)$, hence $\underline{H}^{p'}(F) = \underline{H}^p_Z(F)$.

<u>Proposition 1.3</u>. (Excision Formula) Let Z be locally closed in X, and let V be open in X and such that $Z \subseteq V \subseteq X$. Then for any $F \in C(X)$,

$$H^p_Z(X, F) \cong H^p_Z(V, F|V) \quad .$$

<u>Proof</u>. This follows from the fact that $F \rightsquigarrow \Gamma_Z(X, F)$ and $F \rightsquigarrow \Gamma_Z(V, F|V)$ are isomorphic functors from $C(X)$ to (Ab).

<u>Proposition 1.4</u>. (Spectral Sequence) If Z is locally closed in X, for any F there is a spectral sequence.

$$H^n_Z(X, F) \Leftarrow E_2^{pq} = H^p(X, \underline{H}^q_Z(F)) \quad ,$$

where by $H^p(X, F)$ for any F we mean the right derived functors of the functor $F \rightsquigarrow \Gamma(F)$.

This proposition will follow from the theorem and lemmas below.

<u>Theorem A</u>. Let $\mathcal{C}$, $\mathcal{C}'$, $\mathcal{C}''$ be abelian categories with enough injectives, and suppose given two left-exact covariant functors

$$F: \mathcal{C} \rightarrow \mathcal{C}' \quad , \qquad G: \mathcal{C}' \rightarrow \mathcal{C}'' \quad .$$

Suppose furthermore that F takes injectives into G-acyclic objects, i.e., whenever I is an injective object of $\mathcal{C}$, $R^pG(F(I)) = 0$

for $p > 0$, where R^pG are the right-derived functors of G. Then for each object X in $\mathcal{C}$, there is a spectral sequence relating the right derived functors of F, G, and $G \circ F$.

$$R^n(G \circ F)(X) \Leftarrow E_2^{pq} = R^pG(R^qF(X)) \quad .$$

For the proof of this theorem we refer the reader to [6].

Lemma 1.5. Any injective sheaf in $\mathcal{C}(X)$, the category of abelian sheaves on X, is flasque

Proof. Any sheaf can be embedded in a flasque sheaf, e.g., the sheaf of discontinuous sections of itself. But if I is injective, and $I \subseteq F$ with F flasque, then I is a direct summand of F, hence flasque itself.

Lemma 1.6. If F is flasque, so is $\underline{\Gamma}_Z(F)$.

Proof. Replacing X by an open subset V which contains Z as a closed subset, we may assume that Z is closed in X. Then we must show that if U is any open subset of X, the map

$$\Gamma_Z(X, F) \to \Gamma_{Z \cap U}(U, F|U)$$

is surjective, where we suppose that F is a given flasque sheaf. So let $\sigma \in \Gamma_{Z \cap U}(U, F|U)$, i.e., σ is a section of F over U, whose support is in $Z \cap U$. Consider the zero section of F over the open set $X - Z$. This agrees with σ on the intersection of their domains $U \cap (X - Z) = U - (Z \cap U)$. Hence, there is a section σ' of F over the open set $U \cup (X - Z)$, whose restriction to U is σ, and whose restriction to $X - Z$ is zero. Now since F is flasque, there exists a section σ'' of F over all of X, whose restriction to $U \cup (X - Z)$ is σ'. Clearly σ'' has support in Z, and $\sigma'' \mapsto \sigma$ under the map above. Hence $\underline{\Gamma_Z}(F)$ is flasque.

Lemma 1.7. If F is flasque, then $H^q(X, F) = 0$ for $q > 0$.

Proof. Embed F in an injective sheaf I, and let C be the cokernel:

$$0 \to F \to I \to C \to 0.$$

Then I is flasque by lemma 1.5. Therefore, by [4, Chapter II, 3.1.2], C is also flasque, and we have an exact sequence

$$0 \to \Gamma(F) \to \Gamma(I) \to \Gamma(C) \to 0 \quad .$$

Hence, $H^1(X, F) = 0$, and for $p > 1$,

$$H^p(X, F) \cong H^{p-1}(X, C) = 0$$

by induction on p, since C is also flasque.

Proof of Proposition 1.4. We have the functor $\Gamma_Z : C(X) \to (Ab)$, which may be written as a composite functor

$$\Gamma_Z = \Gamma \bullet \underline{\Gamma_Z}$$

where

$$\underline{\Gamma_Z} : \quad C(X) \to C(X) \quad \text{and} \quad \Gamma : C(X) \to (Ab).$$

In order to apply the theorem quoted above, and thus deduce the existence of the spectral sequence, we need only show that $\underline{\Gamma_Z}$ takes injective sheaves into Γ-acyclic sheaves. But by the lemmas, any injective is flasque, $\underline{\Gamma_Z}$(flasque) = flasque, and any flasque is Γ-acyclic.

Lemma 1.8. Let Z be locally closed in X, let Z' be closed in Z, and let $Z'' = Z - Z'$. Then for any abelian sheaf F on X, there is an exact sequence

$$0 \to \Gamma_{Z'}(F) \to \Gamma_Z(F) \to \Gamma_{Z''}(F) \quad .$$

Moreover, if F is flasque, we can write a 0 on the right also.

Proof. Clearly we may assume that Z is closed in X. Then $\Gamma_{Z'}(F)$ is the set of those global sections of F with support in Z', which clearly contained in $\Gamma_Z(F)$. Letting $V = \complement Z'$ in X, V is open, and $Z'' = Z \cap V$, i.e., Z'' is closed in V. Then $\Gamma_{Z''}(F)$ = the set of elements $\sigma \in \Gamma(V, F|V)$ whose support is in Z''. Now it is clear that the natural restriction map

$$\phi : \Gamma(X,F) \to \Gamma(V, F|V)$$

induces a map $\phi_Z : \Gamma_Z(X,F) \to \Gamma_{Z''}(X,F)$. Furtherfore, to say $\phi_Z(\sigma) = 0$ is to say σ is zero on Z'', i.e., σ has support in Z'. Hence, our sequence is exact.

Now if F is flasque, ϕ is surjective. Hence, if $\sigma \in \Gamma_{Z''}(X,F)$, $\exists\, \sigma' \in \Gamma(X,F)$ restricting to σ. But since $V = \complement Z'$, σ' must have support in Z'. Hence, ϕ_Z is surjective.

<u>Proposition 1.9</u>. Let Z, Z', Z'' be as in the lemma, and let F be any abelian sheaf on X. Then there are exact sequences,

$$0 \to \Gamma_{Z'}(X,F) \to \Gamma_Z(X,F) \to \Gamma_{Z''}(X,F) \to H^1_{Z'}(X,F) \to H^1_Z(X,F) \to \dots$$

and

$$0 \to \underline{\Gamma}_{Z'}(F) \to \underline{\Gamma}_Z(F) \to \underline{\Gamma}_{Z''}(F) \to \underline{H}^1_{Z'}(F) \to \underline{H}^1_Z(F) \to \dots$$

<u>Corollary 1.9</u>. If Z is closed in X, and F is any abelian sheaf on X, there are exact sequences

$$0 \to \Gamma_Z(X,F) \to \Gamma(X,F) \to \Gamma(X-Z,F) \to H^1_Z(X,F) \to$$
$$\to H^1(X,F) \to H^1(X-Z,F) \to \dots$$

and

$$0 \to \underline{\Gamma}_Z(F) \to F \to j_*(F|X-Z) \to \underline{H}^1_Z(F) \to 0$$

$$\underline{H}^{p+1}_Z(F) \cong R^p j_*(F|X-Z) \qquad \text{for } p > 0 \quad ,$$

where $j : X - Z \hookrightarrow X$ is the natural injection, and j_* is the direct image functor.

Proofs. Take an injective resolution $\mathcal{I} = (I_i)$ of F:

$$0 \to F \to I_0 \to I_1 \to \dots \quad .$$

Then the I_i are all flasque by lemma 1.5, and so by lemma 1.8 we have an exact sequence of complexes

$$0 \to \Gamma_{Z'}(\mathcal{I}) \to \Gamma_Z(\mathcal{I}) \to \Gamma_{Z''}(\mathcal{I}) \to 0 \quad ,$$

and similarly for the sheaf functors $\underline{\Gamma}_{Z'}$, $\underline{\Gamma}_Z$, $\underline{\Gamma}_{Z''}$. These complexes give rise to the exact sequences written above.

For the corollary, we take the particular case where $Z = X$, and write Z for Z'. Note that $\underline{\Gamma}_X(F) = F$, and $\underline{\Gamma}_{X-Z}(F) = j_*(F|X-Z)$ by the definition of the direct image functor. $\underline{\Gamma}_X$ is exact, so $\underline{H}^p_X(F) = 0$ for $p > 0$, and $\underline{H}^p_{X-Z}(F) = R^p j_*(F|X-Z)$.

Proposition 1.10. Let F be an abelian sheaf on X. Then the following conditions are equivalent:

(i) F is flasque.

(ii) For all locally closed subspaces Y of X, and for all $i > 0$,

$$H^i_Y(X, F) = 0 \quad \text{and} \quad \underline{H}^i_Y(F) = 0 \quad .$$

(iii) For all closed subspaces Y of X, $H^1_Y(X,F) = 0$.

Proof. (i) => (ii). Suppose F flasque, and $Y \subseteq X$ locally closed. Let $V \subseteq X$ be an open set containing Y as a closed subset. Then $F|V$ is flasque, so by the excision formula (proposition 1.3), we may assume that $V = X$, i.e., Y is closed in X. Now using the corollary of Proposition 1.9, and noting that $H^i(X,F) = H^i(X - Y, F) = 0$ for $i > 0$, since F is flasque, we find that $H^i_Y(X,F) = 0$ for $i \geq 2$, and

$$\Gamma(X,F) \to \Gamma(X - Y, F) \to H^1_Y(X,F) \to 0 \quad .$$

By the definition of flasqueness, also $H^1_Y(X,F) = 0$. For any open $U \subseteq X$, $F|U$ is also flasque, so $H^i_{Y\cap U}(U, F|U) = 0$ for $i > 0$. Passing to associated sheaves (Proposition 1.2), we find that $\underline{H}^i_Y(F) = 0$ for $i > 0$.

(ii) => (iii) is trivial.

(iii) => (i) follows from the definition of flasqueness and the exact sequence, for any closed Y,

$$\Gamma(X,F) \to \Gamma(X - Y, F) \to H^1_Y(X,F) \quad .$$

Proposition 1.11. Let Y be a closed subset of X, let $V = X - Y$, let F be an abelian sheaf on X, and let n be an integer. Then the following conditions are equivalent:

(i) For all $i \leq n$, $\underline{H}^i_Y(F) = 0$.

(ii) For all open subsets U of X, the map $\alpha_i : H^i(U, F) \to H^i(U \cap V, F)$ is injective for $i = 0$, and an isomorphism for all $i < n$.

Proof. (i) => (ii). Since the condition (i) is of a local nature, it will be sufficient to prove (ii) in the case where $U = X$. So we must show that the map

$$\alpha_i : H^i(X, F) \to H^i(X - Y, F)$$

is injective for $i = 0$, and an isomorphism for $i < n$. Using the exact sequence of the Corollary to Proposition 1.9, it will be sufficient to show that $H^i_Y(X, F) = 0$ for $i \leq n$. But by Proposition 1.4 there is a spectral sequence

$$H^i_Y(X, F) <= E_2^{pq} = H^p(X, \underline{H}^q_Y(F)) \quad .$$

Since by hypothesis the $\underline{H}^q_Y(F) = 0$ for $q \leq n$, it follows that the abutment $H^i_Y(X, F) = 0$ for $i \leq n$.

(ii) => (i). Our hypothesis clearly insures that $H^i_{Y \cap U}(U, F|U) = 0$ for any open U, and for $i = 0$ and $i < n$. Hence, passing to associated sheaves, $\underline{H}^i_Y(F) = 0$ for $i = 0$ and $i < n$. The only remaining case is where $i = n > 0$. Then for each U, we have

$$H^{n-1}(U,F) \xrightarrow{\sim} H^{n-1}(U \cap V, F) \to H^n_{Y\cap U}(U,F) \to H^n(U,F) .$$

This exact sequence commutes with the restriction homomorphisms as we pass from one U to another, hence gives rise to an exact sequence of associated sheaves. But the sheaf associated to the presheaf $U \rightsquigarrow H^n(U,F)$ is zero, since $n > 0$! So we have

$$\underline{H}^{n-1}(F) \xrightarrow{\sim} R^{n-1}j_*(F|V) \to \underline{H}^n_Y(F) \to 0$$

which shows $\underline{H}^n_Y(F) = 0$.

Proposition 1.12. Let X be a Zariski space of dimension n, let Y be a locally closed subset of X, and let F be any abelian sheaf on X. Then for all $i > n$, $H^i_Y(X,F) = 0$, and $\underline{H}^i_Y(F) = 0$.

Proof. See [6, § 3.6] and [4, II, Th. 4.15.2], where the analogous theorem is proved for $H^i(X,F)$. The same proof works also for $H^i_Y(X,F)$, once one observes that $H^i_Y(X,F) = 0$ for $i > 0$ and F flasque (Proposition 1.10), and that the functor Γ_Y commutes with direct limits and finite direct sums. Any open subset $U \subseteq X$ is again a Zariski space of dim n, so $H^i_{Y\cap U}(U, F|U) = 0$ for $i > n$. Passing to associated sheaves gives $\underline{H}^i_Y(F) = 0$ for $i > n$.

Remark. For suitably nice topological spaces in which sheaf cohomology agrees with singular cohomology (e.g., paracompact and locally contractible)* one may interpret the cohomology groups defined above in terms of known relative cohomology. To be precise, let Y be closed in X and let G be an abelian group (respectively the constant sheaf G on X). Then

$$H^i_Y(X,G) = H^i(X, X-Y;G) \quad ,$$

where the expression the right is singular cohomology of X relative to $X - Y$, with coefficients in G. This is a canonical, functorial isomorphism, compatible with the usual exact sequences.

Example. Let X be a topological space, and Y a closed subset, so that X looks locally like $Y \times \mathbb{R}^d$: That is, such that for each $y \in Y$, there is a neighborhood V of y in X, such that the pair $(Y \cap V, V)$ is homeomorphic to $(Y \cap V \times \{0\}, Y \cap V \times \mathbb{R}^d)$.

We consider the cohomology sheaf $\underline{H}^i_Y(X,\mathbb{Z})$. This sheaf will have support in Y, and we can calculate it locally. Over a V such as above, it looks like

$$H^i_{\{0\}}(\mathbb{R}^d, \mathbb{Z}) = \begin{cases} 0 & \text{if } i \neq d \\ \mathbb{Z} & \text{if } i = d \end{cases} .$$

(One can calculate this by ordinary topological methods via the remark above: The pair $(\mathbb{R}^d, \mathbb{R}^d - \{0\})$ is of the same homotopy type as the pair $(\mathbb{R}^d, S^{d-1})$, so we have

*see [14, exposé XIII p.3] and [15, exposé XX, p.1]. Compare also [4,II5.10.1] which says that these also agree with Čech cohomology.

$$\cdots \to H^{i-1}(\mathbb{R}^d; \mathbb{Z}) \to H^{i-1}(S^{d-1}; \mathbb{Z}) \to H^i_{\{0\}}(\mathbb{R}^d, \mathbb{Z}) \to H^i(\mathbb{R}^d, \mathbb{Z}) \to \cdots$$

We know $H^i(\mathbb{R}^d; \mathbb{Z}) = 0$ for $i > 0$, and $H^i(S^{d-1}; \mathbb{Z}) = 0$ if $i \neq d - 1$, $\mathbb{Z}$ otherwise.)

Hence the sheaf $\underline{H}^d_Y(X, \mathbb{Z})$ is locally isomorphic to $\mathbb{Z}$ along Y. Call this sheaf $\underline{\mathbb{T}_{Y,X}}$, the <u>sheaf of twisted integers of</u> Y <u>in</u> X. The spectral sequence of Proposition 1.4 degenerates, since $\underline{H}^i_Y(X, \mathbb{Z}) \neq 0$ only when $i = d$, and we find

$$H^i_Y(X, \mathbb{Z}) \cong H^{i-d}(X, \mathbb{T}_{Y,X}) = H^{i-d}(Y, \mathbb{T}_{Y,X}) \quad ,$$

since $\mathbb{T}_{Y,X}$ is concentrated on Y. Substituting in the exact sequence of Proposition 1.9, we have

$$\cdots \to H^i(X, \mathbb{Z}) \to H^i(X - Y, \mathbb{Z}) \to H^{i+1-d}(Y, \mathbb{T}_{Y,X}) \to H^{i+1}(X, \mathbb{Z}) \to \cdots .$$

§ 2. Application of Local Cohomology to Preschemes.

In this section we will apply the notions of local cohomology, as developed in Section 1, to the case where our topological space X is a prescheme. We will first study the case of an affine scheme, showing how to compute the local cohomology groups by means of Koszul complexes. Then we will give a theorem valid on quite general preschemes, which interprets the local cohomology groups and sheaves in terms of the "Ext" functors. These results are very closely related to theorems of Serre [10, § 69] relating cohomology of sheaves on projective space to "Ext" groups.

Proposition 2.1. Let X be a prescheme, with structure sheaf $\mathcal{O}_X$, and let $Y = V - U$ be a locally closed subset, where U and V are open. Assume that the natural injection maps of U and V into X are quasi-compact immersions. (This will always be true if X is locally Noetherian.) If F is a quasi-coherent sheaf of $\mathcal{O}_X$-modules, then the sheaves $\underline{H}_Y^n(F)$ are quasi-coherent sheaves of $\mathcal{O}_X$-modules, for all $n \geq 0$.

Proof. By Proposition 1.1 we have an exact sequence of sheaves

$$\ldots \to \underline{H}_U^n(F) \to \underline{H}_Y^n(F) \to \underline{H}_V^{n+1}(F) \to \ldots .$$

Since the kernel and cokernel of a homomorphism of quasi-coherent sheaves on any prescheme are quasi-coherent, and since an extension of one quasi-coherent sheaf by another is quasi-coherent*, it will be sufficient to show that the sheaves $\underline{H}^n_U(F)$ and $\underline{H}^n_V(F)$ are quasi-coherent. We have already seen (Corollary 1.9) that if U is open in X, and $j : U \to X$ is the natural injection, then the sheaf $\underline{H}^n_U(F)$ is just $R^n j_*(F|U)$. Since $F|U$ is a quasi-coherent sheaf on U, and since an immersion is separated, our proposition will result from the following theorem, which we suppose known.

Theorem B. Let $f : X \to Y$ be a quasi-compact separated morphism of preschemes, and let F be a quasi-coherent sheaf on X. Then the higher direct images $R^n f_*(F)$ of F are all quasi-coherent sheaves on Y.

Proof. See [5, I.1.6.3 and III 1.4.10]

Proposition 2.2. Let $X = \operatorname{Spec} A$ be an affine scheme, and let Y be a closed subset of X. Then for any quasi-coherent sheaf F on X, and for any $i \geq 0$, $\underline{H}^i_Y(F)$ is the sheaf associated to the A-module $H^i_Y(X, F)$. Moreover, there is an exact sequence

$$0 \to H^0_Y(X, F) \to H^0(X, F) \to H^0(X-Y, F) \to H^1_Y(X, F) \to 0$$

*Since the question is local, it is sufficient to consider an affine scheme. Then the first statement follows from [5, I. 1.3.9] and [5, I.1.4.1], and the second statement is [5, III. 1.4.17].

and there are isomorphisms

$$H^i(X-Y, F) \xrightarrow{\sim} H_Y^{i+1}(X, F) \qquad i > 0 .$$

Proof. For the first statement, we apply the spectral sequence (Proposition 1.4)

$$H_Y^n(X, F) \Leftarrow E_2^{pq} = H^p(X, \underline{H}_Y^q(F)) .$$

By the previous proposition, we know that the sheaves $\underline{H}_Y^q(F)$ are quasi-coherent. Therefore, since X is affine, $E_2^{pq} = 0$ for $p > 0$. So our spectral sequence degenerates and we have

$$H_Y^n(X, F) \cong H^0(X, \underline{H}_Y^n(F)) ,$$

or, in other words, $\underline{H}_Y^n(F)$ is the sheaf associated to the A-module $H_Y^n(X, F)$.

The second statement follows directly from the exact sequence of Corollary 1.9, since for X affine, the groups $H^i(X, F) = 0$ for $i > 0$.

We now proceed to a detailed study of the local cohomology of an affine scheme. Since the cohomology sheaves $\underline{H}_Y^n(F)$ are well expressed in terms of the groups $H_Y^n(X, F)$ by the above proposition, we will restrict our attention to the latter.

We first recall the definition of the Koszul complex (see [11, Ch. IV] or [5, Ch. III, § 1.1]). Let A be a commutative ring, and let f be an element of A. Then we denote by $K(f)$ the (homological) complex of A-modules defined as follows: The modules are

$$K_1(f) \cong K_0(f) \cong A \quad ; \qquad K_i(f) = 0 \quad \text{for} \quad i > 1 \quad ,$$

and the map

$$d : K_1(f) \to K_0(f)$$

is multiplication by f.

If $\underline{f} = (f_1, \ldots, f_n)$ is a finite family of elements of A, and if M is an A-module, then we denote by $K_*(\underline{f};M)$ the homological complex

$$K(f_1) \otimes_A \cdots \otimes_A K(f_n) \otimes_A M \quad ,$$

where M is considered as a complex concentrated in degree zero. We denote by $K^*(\underline{f};M)$ the cohomological complex

$$\mathrm{Hom}_A(K_*(\underline{f}; A), M) \quad .$$

The homology and cohomology of these complexes will be denoted by $H_*(\underline{f}; M)$ and $H^*(\underline{f}; M)$, respectively.

If m and m' are two positive integers such that $m' \geq m$, then there is a natural map of the Koszul complexes

$$K(f^{m'}) \to K(f^m)$$

defined as follows: In degree zero it is the identity map of A, and in degree one it is multiplication by $f^{m'-m}$. One extends this definition in the obvious way to give maps of the Koszul complexes of n elements of A with respect to a module M, and therefore also maps of the homology and cohomology of these complexes, as follows (letting $\underline{f}^m = (f_1^m, \ldots, f_n^m)$) :

$$K_*(\underline{f}^{m'};M) \to K_*(\underline{f}^m;M)$$

$$H_*(\underline{f}^{m'};M) \to H_*(\underline{f}^m;M)$$

and

$$K^*(\underline{f}^{m'};M) \leftarrow K^*(\underline{f}^m;M)$$

$$H^*(\underline{f}^{m'};M) \leftarrow H^*(\underline{f}^m;M) .$$

Thus in the first case we have inverse systems of complexes and homology groups; in the second case we have direct systems of complexes and cohomology groups.

Theorem 2.3. Let A be a commutative ring with prime spectrum X; let $\underline{f} = (f_1, \ldots, f_n)$ be a finite family of elements of A, and let the variety of the ideal they generate be $Y \subseteq X$; let M be an A-module. Then there is an isomorphism of cohomological functors $(i \geq 0)$

$$\lim_{\overrightarrow{m}} H^i(\underline{f}^m;M) \cong H^i_Y(X, \tilde{M}) \quad .$$

The complete proof of this theorem is beyond the scope of these notes, and besides, is fairly well known. We will content ourselves with outlining the proof, and refer to the literature for details. [5, Ch. III]

The first step is to interpret the direct limit of Koszul cohomology in terms of Cech cohomology. For simplicity of notation, let us define $H^i_{\underline{f}}(M)$ to be $\lim_{\overrightarrow{m}} H^i(\underline{f}^m;M)$. If $\mathfrak{U}$ is a collection of open sets of X, and if F is a sheaf on X, we will denote by $\check{H}^i(\mathfrak{U}, F)$ the ith Cech cohomology group of X with respect to the family of open sets $\mathfrak{U}$, and with coefficients in the sheaf F. Finally, if X = Spec A, and if $\underline{f} = (f_1, \ldots, f_n)$ is a finite family of elements of A, then we will denote by $\mathfrak{U}_{\underline{f}}$ the family of open sets (U_i), $i = 1, \ldots, n$, where U_i is the complement in X of the variety of the ideal generated by f_i.

Proposition C. [5, Ch. III, 1.2.3] With the hypotheses of Theorem 2.3, and with the notations above, there is an exact sequence

$$0 \to H^0_{\underline{f}}(M) \to M \xrightarrow{\alpha} \check{H}^0(\mathfrak{U}_{\underline{f}}, \tilde{M}) \to H^1_{\underline{f}}(M) \to 0 \quad ,$$

where α is the natural restriction map and there are isomorphisms

$$H^i(\mathcal{U}_{\underline{f}}, \tilde{M}) \cong H^{i+1}_{\underline{f}}(M) \qquad \text{for } i > 0 .$$

Moreover, the exact sequence and isomorphisms are functorial in M.

The proof of this proposition is elementary. One shows explicitly that the complex $\mathcal{C}^*(\mathcal{U}_{\underline{f}}, \tilde{M})$ of Cech cochains of $\mathcal{U}_{\underline{f}}$ with coefficients in $\tilde{M}$ is canonically isomorphic to the direct limit $\mathcal{K}^*_{\underline{f}}(M)$ of the cohomological Koszul complexes $K^*(\underline{f}^m; M)$, but with the dimensions shifted by one. That is to say, there is a commutative diagram

$$\begin{array}{ccccccc} 0 \to \mathcal{K}^0_{\underline{f}}(M) & \to & \mathcal{K}^1_{\underline{f}}(M) & \to & \mathcal{K}^2_{\underline{f}}(M) & \to & \cdots \\ & & \downarrow\approx & & \downarrow\approx & & \cdots \\ & 0 \to & \mathcal{C}^0(\mathcal{U}_{\underline{f}}, \tilde{M}) & \to & \mathcal{C}^1(\mathcal{U}_{\underline{f}}, \tilde{M}) & \to & \cdots \end{array}$$

Now since taking direct limits is an exact functor, one can compute $H^i_{\underline{f}}(M)$ as the cohomology of the complex $\mathcal{K}^*_{\underline{f}}(M)$, and one finds the exact sequence and isomorphisms of the proposition. The entire proof is functorial in M, so the resulting exact sequence and isomorphisms are also.

Theorem D. Let X be a scheme, let $\mathcal{U}$ be a cover of X by open affine subschemes, and let F be a quasi-coherent sheaf on X. Then there is an isomorphism of cohomological functors $(i \geq 0)$

$$\check{H}^i(\mathfrak{U}, F) \cong H^i(X, F) \quad ,$$

where $H^i(X, F)$ denotes the ith right derived functor of the Γ-functor from the category of all abelian sheaves on X.

For the proof see [5, Ch. III, Prop. 1.4.1].

This theorem allows us to replace $H^i(\mathfrak{U}_{\underline{f}}, \tilde{M})$ by $H^i(X-Y, \tilde{M})$ in the previous proposition, since $\mathfrak{U}_{\underline{f}}$ is a cover of $X - Y$ by open affines. Now with the aid of Proposition 2.2, we can complete the proof of Theorem 2.3. For the functors $H^i_Y(X, \tilde{M})$ and $H^i_{\underline{f}}(M)$ have both been characterized in the same way, namely, as the kernel and cokernel of the map $\alpha : M \to \Gamma(X - Y, \tilde{M})$ for $i = 0, 1$, and as $H^{i-1}(X - Y, \tilde{M})$ for $i \geq 2$. Hence they are isomorphic functors for each i. That the isomorphism commutes with connecting homomorphisms can be seen by going back to the isomorphism of cochain complexes described in the proof of Proposition C.

Our next goal will be to show if A is a Noetherian ring, then the functors $H^i_{\underline{f}}(M)$ for $i > 0$ are the right derived functors of the functor $H^0_{\underline{f}}(M)$ in the category of A-modules. It is the same thing to say if M is an injective A-module, then $H^i_{\underline{f}}(M) = 0$ for $i > 0$.

Definition. Let $(M_m)_{m \geq 0}$ be an inverse system of abelian groups. We say it is essentially zero if for each $m \geq 0$, there exists an $m' \geq m$ such that the map

$$M_{m'} \to M_m$$

is the zero map.

Remark. 1) If the inverse system (M_m) is essentially zero, then it follows that its inverse limit $\varprojlim M_m$ is zero. The converse is false.

2) If we have an exact sequence of inverse systems

$$0 \to (M'_m) \to (M_m) \to (M''_m) \to 0 \quad ,$$

then the middle one is essentially zero if and only if the two outside ones are essentially zero. We leave the proof as an easy exercise for the reader.

Lemma 2.4. Let A be a commutative ring, let $\underline{f} = (f_1, \ldots, f_n)$ be a finite family of elements of A, and let $i > 0$ be an integer. Then the following statements are equivalent:

(i) $H^i_{\underline{f}}(M) = 0$ for all injective A-modules M.

(ii) $(H_i(\underline{f}^m; A))_{m \geq 1}$ is an essentially zero inverse system.

Proof. By definition,

$$H^i_{\underline{f}}(M) = \varinjlim_m H^i(\underline{f}^m; M) \quad .$$

If M is injective, then $\operatorname{Hom}(\quad, M)$ is exact. Hence it commutes with passage to homology, and we have

$$H^i(\underline{f}^m; M) \cong \operatorname{Hom}_A(H_i(\underline{f}^m; A), M) \quad .$$

If $(H_i(\underline{f}^m; A))_{m \geq 1}$ is essentially zero, then for each m there is an $m' \geq m$ such that

$$H^i(\underline{f}^m;M) \to H^i(\underline{f}^{m'};M)$$

is zero, and hence the direct limit $H^i_{\underline{f}}(M)$ of these modules is zero.

Conversely, suppose (i) is true. Given an integer $m > 0$, we can imbed $H_i(\underline{f}^m;A)$ in an injective module M. Let

$$\alpha \in \mathrm{Hom}_A(H_i(\underline{f}^m;A), M) \cong H^i(\underline{f}^m;M)$$

be the imbedding map. If the direct limit $H^i_{\underline{f}}(M)$ is zero, then there must be an $m' \geq m$ such that the image of α in $H^i(\underline{f}^{m'};M)$ is zero; in other words, such that the composed map

$$H_i(\underline{f}^{m'};A) \to H_i(\underline{f}^m;A) \xrightarrow{\alpha} M$$

is zero. Since α is a monomorphism, it follows that the first arrow is zero.

Lemma 2.5. Let A be a Noetherian ring, let $\underline{f} = (f_1, \ldots, f_n)$ be a finite family of elements of A, and let N be an A-module of finite type. Then the inverse system of A-modules

$$(H_i(f^m;N))_{m \geq 1}$$

is essentially zero for $i > 0$.

Proof. We proceed by induction on n. First suppose $n = 1$. Then the only value of i to consider is $i = 1$. One sees immediately that $H_1(f^m;N)$ is the submodule N_m of N consisting of those elements annihilated by f^m. The map $N_{m'} \to N_m$, for $m' \geq m$, is multiplication by $f^{m'-m}$. Now the submodules N_m of N form an increasing sequence, which must be stationary, since A is Noetherian and N of finite type. In other words, there is an m_0 such that f^{m_0} annihilates all of the modules N_m. Therefore, if m is given, and $m' = m + m_0$, the map from $N_{m'} \to N_m$ is multiplication by f^{m_0}, i.e., the zero map. Hence the inductive system (N_m) is essentially zero.

Now let n be an integer ≥ 2, and suppose the lemma proved for all sequences of $< n$ elements of A, and for all modules N of finite type. Given $\underline{f} = (f_1, \ldots, f_n) \varepsilon A$, and a module of finite type N, let $\underline{g} = (f_1, \ldots, f_{n-1})$. Then for each $i \geq 0$ there is an exact sequence [11, Ch. IV],

$$0 \to H_0(f_n^m;H_i(\underline{g}^m;N)) \to H_i(\underline{f}^m;N) \to H_1(f_n^m;H_{i-1}(\underline{g}^m;N)) \to 0$$

To show that the middle inverse system is essentially zero, it will be sufficient, by the remark 2 above, to show that the two outside inverse systems are essentially zero.

On the left, we factor the inverse system map as follows (where $m' \geq m$):

$$\begin{array}{ccc} H_0(f_n^{m'};H_i(g^{m'};N)) & \xrightarrow{\ \alpha\ } & H_0(f_n^{m'};H_i(g^{m};N)) \\ \downarrow & \swarrow_{\beta} & \\ H_0(f_n^{m};H_i(g^{m};N)) & & \end{array}$$

By the induction hypothesis, the inverse system $(H_i(g^m;N))_m$ is essentially zero for $i > 0$. Therefore, when m' is enough greater than m, the map α will be zero, and hence also $\beta \circ \alpha$. Therefore, the left-hand inverse system is essentially zero.

We perform a similar factorization on the right-hand inverse system.

$$\begin{array}{ccc} H_1(f_n^{m'};H_{i-1}(g^{m'};N)) & \xrightarrow{\ \alpha\ } & H_1(f_n^{m'};H_{i-1}(g^{m};N)) \\ \downarrow & \swarrow_{\beta} & \\ H_1(f_n^{m};H_{i-1}(g^{m};N)) & & \end{array}$$

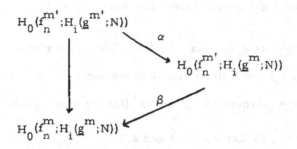

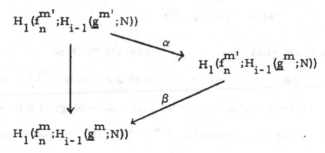

For given i, m, N, the A-module $H_{i-1}(g^m;N)$ is of finite type. Therefore, applying the induction hypothesis to this module, and to the single element $f_n \in A$, we find that for m' enough larger than m, the map β is zero. Hence, $\beta \circ \alpha$ is zero, and the inverse system on the right is also essentially zero. Q.E.D.

Proposition 2.6. Let A be a Noetherian ring, let $\underline{f} = (f_1, \ldots, f_n)$ be a finite family of elements of A, and let M be an injective A-module. Then $H^i_{\underline{f}}(M) = 0$ for $i > 0$.

Proof. By Lemmas 2.4 and 2.5.

Corollary 2.7. Let A be a Noetherian ring with prime spectrum X, and let M be an injective A-module. Then the sheaf $\tilde{M}$ on X is a flasque sheaf.

Proof. To show that $\tilde{M}$ is flasque, we must show that for any closed $Y \subseteq X$, the map

$$M \xrightarrow{\alpha} \Gamma(X - Y, \tilde{M})$$

is surjective. Let $f_1, \ldots, f_n$ generate the ideal of Y. Then $X - Y$ is the union of the open sets of the family $\mathcal{U}_{\underline{f}}$ defined above, and from the definition of Cech cohomology it follows that $\Gamma(X-Y, \tilde{M})$ is isomorphic to $\check{H}^0(\mathcal{U}_{\underline{f}}, \tilde{M})$. So by Proposition C quoted above, there is an exact sequence

$$M \xrightarrow{\alpha} \Gamma(X - Y, \tilde{M}) \to H^1_{\underline{f}}(M) \to 0 \quad .$$

But $H^1_{\underline{f}}(M) = 0$ for M injective by the previous proposition, so α is surjective.

Remarks. Conversely, the corollary implies the proposition as we see by Proposition C above. Using results on the structure of injective modules over Noetherian rings [3], [9]* one can give a direct proof of the corollary as follows. An injective over a Noetherian ring A is a direct sum of injective hulls I_p of residue class fields $k(p)$ of prime ideals p of A. Since a direct sum of flasque sheaves is flasque, it suffices to show that $\tilde{I}_p$ is flasque for each p. But I_p is a direct limit of Artin modules over the local ring A_p. Hence $\tilde{I}_p$ is a constant sheaf I_p on the closed subset $V(p)$ of Spec A, and zero outside. Since a constant sheaf on an irreducible Noetherian space is flasque, the contention follows.

Now we come to the main theorem of this section, relating the local cohomology groups on a prescheme to a direct limit of Exts.

Let X be a prescheme, let Y be a closed subspace of X, and let F be a quasi-coherent sheaf of $\mathcal{O}_X$-modules. Let $\mathcal{J}$ be a quasi-coherent sheaf of ideals defining Y, and for each $n \geq 1$, let $\mathcal{O}_n = \mathcal{O}_X / \mathcal{J}^n$. Then $\mathcal{O}_n$ is a sheaf concentrated on Y, and for each n there is a natural injection

$$\mathrm{Hom}_{\mathcal{O}_X}(\mathcal{O}_n, F) \to \Gamma_Y(X, F) .$$

For a homomorphism of $\mathcal{O}_n$ into F is determined by the image of the unit section of $\mathcal{O}_n$, which must be a section of F with support in Y.

---

*See also [17, II § 7] for a complete description of the structure of injective $\mathcal{O}_X$- modules on a locally noetherian prescheme X. Using these results allows one to avoid the painful (2.3) - (2.6). It also gives a more direct proof of Proposition 2.1 for X locally noetherian.

Letting F range over the category of $\mathcal{O}_X$-modules, we consider on the one hand the derived functors $\mathrm{Ext}^i_{\mathcal{O}_X}(\mathcal{O}_n, F)$ of $\mathrm{Hom}_{\mathcal{O}_X}(\mathcal{O}_n, F)$, and on the other hand the cohomological functor $H^i_Y(X, F)$, which is a priori not a derived functor in this category!* Having a map of these functors for $i = 0$, we deduce one for $i > 0$ by the universal property of derived functors.

$$\mathrm{Ext}^i_{\mathcal{O}_X}(\mathcal{O}_n, F) \to H^i_Y(X, F) .$$

As n varies, these Ext's form a direct system mapping into $H^i_Y(X, F)$, so there are homomorphisms

$$\varinjlim_n \mathrm{Ext}^i_{\mathcal{O}_X}(\mathcal{O}_n, F) \to H^i_Y(X, F) . \tag{*}$$

Performing these homomorphisms locally, and passing to associated sheaves, we have homomorphisms of sheaves

$$\varinjlim_n \underline{\mathrm{Ext}}^i_{\mathcal{O}_X}(\mathcal{O}_n, F) \to \underline{H}^i_Y(F) . \tag{$\underline{*}$}$$

Theorem 2.8. If X is locally Noetherian, and F quasi-coherent, then the homomorphisms ($\underline{*}$) are isomorphisms. If furthermore X is Noetherian, then the homomorphisms (*) are also isomorphisms.

The proof will be deferred until after some auxialiary propositions and lemmas.

---

*In fact, it is a derived functor, because by lemma 2.9 below, any injective $\mathcal{O}_X$-module is flasque, and by Proposition 1.10, flasque sheaves are Γ_Y-acyclic, so may be used to calculate the cohomology groups $H^i_Y(X, F)$.

[4, II, 7.3.2]

Lemma 2.9. Let X be a prescheme, let F and G be sheaves of $\mathcal{O}_X$-modules, and suppose that G is injective in the category of $\mathcal{O}_X$-modules. Then $\underline{\mathrm{Hom}}_{\mathcal{O}_X}(F,G)$ is a flasque sheaf.

Proof. We must show, if U is open in X, then the natural restriction

$$\mathrm{Hom}_{\mathcal{O}_X}(F,G) \to \mathrm{Hom}_{\mathcal{O}_U}(F|U, G|U)$$

is surjective. Let F_U denote the unique sheaf whose restriction to U is $F|U$, and which is zero outside of U. Then a homomorphism of $F|U$ into $G|U$ is the same as a homomorphism of F_U into G. Since $F_U \subseteq F$, and since G is injective, any such homomorphism extends to a homomorphism of F into G.

[4, II 7.3.3]

Proposition 2.10. Let X be a prescheme, and let F and G be sheaves of $\mathcal{O}_X$-modules. Then there is a spectral sequence

$$\mathrm{Ext}^n_{\mathcal{O}_X}(F,G) \Leftarrow E_2^{pq} = H^p(X, \underline{\mathrm{Ext}}^q_{\mathcal{O}_X}(F,G)) \quad .$$

Proof. One can represent $\mathrm{Hom}_{\mathcal{O}_X}(F,G)$ as a composite functor

$$\mathrm{Hom}_{\mathcal{O}_X}(F,G) = \Gamma(\underline{\mathrm{Hom}}_{\mathcal{O}_X}(F,G)) \quad .$$

By Lemma 2.9 $Hom_{\mathcal{O}_X}(F,G)$, considered as a functor in G, takes injectives into Γ-acyclic objects (Lemma 1.7). Therefore, we may apply Theorem A quoted above to deduce the existence of the spectral sequence of derived functors.

[4, ch.II, § 4.12]

Proposition 2.11. Let X be a Noetherian topological space, and let (F_i) be a direct system of abelian sheaves on X. Then for each $p \geq 0$,

$$H^p(X, \varinjlim F_i) = \varinjlim H^p(X, F_i) \quad .$$

Proof. Let $F = \varinjlim F_i$. To calculate the cohomology of the F_i and of F, take injective resolutions $C(F_i)$ of the F_i, in such a way that they form a direct system, and let $C(F) = \varinjlim C(F_i)$. Then $C(F)$ is a resolution of F by sheaves which are direct limits of injectives. Using the quasi-compacity of a Noetherian space, one finds that the Γ functor commutes with direct limits. Now since the direct limit functor is exact, we have

$$\varinjlim \quad H^p(X, F_i) = H^p(\Gamma(C(F))) \quad .$$

It will therefore be sufficient to show that any direct limit of injective sheaves is flasque, since then we may use the complex $C(F)$ to calculate the cohomology of F. But by Lemma 1.5, any injective sheaf is flasque, and an easy argument using quasi-compacity shows that a direct limit of flasque sheaves on a Noetherian space is flasque.

Example. The previous proposition is false without the Noetherian hypothesis. For example, let X be the space of positive integers with the discrete topology. Let F_i be the sheaf whose stalk is $\mathbb{Z}$ for $n \geq i$, and 0 for $n < i$. Map F_i into F_{i+1} by killing the ith stalk, and leaving everything else alone. Then $\varinjlim F_i = 0$, but

$$\varinjlim \Gamma(F_i) = \prod_{n \in X} \mathbb{Z} / \bigoplus_{n \in X} \mathbb{Z} \neq 0 .$$

Lemma 2.12. Let A be a Noetherian ring with prime spectrum X. Let M, N be two A-modules and let M be of finite type. Then there are canonical functorial isomorphisms

$$\operatorname{Ext}^i_A(M,N)^\sim \cong \underline{\operatorname{Ext}}^i_{\mathcal{O}_X}(\tilde{M}, \tilde{N}) ,$$

where a tilde denotes taking the quasi-coherent sheaf on X associated to the given A-module.

Proof. For $i = 0$, the isomorphism follows from the fact that $\operatorname{Hom}(M,N)$ commutes with localization when M is of finite presentation [5, I, 1.3.12 (ii)]. As M varies in the category $\mathcal{C}^f_A$ of modules of finite type over A, $\operatorname{Ext}^i_A(M,N)^\sim$ is the ith right derived functor (with respect to the first variable) of $\operatorname{Hom}_A(M,N)^\sim$. Hence, there are canonical homomorphisms (of functors in M, from $\mathcal{C}^f_A$ to the category of sheaves of $\mathcal{O}_X$-modules)

$$\mathrm{Ext}^i_A(M,N)^\sim \to \underline{\mathrm{Ext}}^i_{\mathcal{O}_X}(\tilde{M}, \tilde{N}) \qquad (1)$$

Since A is Noetherian, every $M \in \mathcal{C}^f_A$ has a projective resolution by finitely generated free A-modules. Hence to show (1) is an isomorphism, we need only show the right-hand side is zero when $M = A^r$, for then they will be derived functors. Since $\underline{\mathrm{Ext}}^i_{\mathcal{O}_X}$ Commutes with finite direct sums, it is sufficient to show that $\underline{\mathrm{Ext}}^i_{\mathcal{O}_X}(\tilde{A}, \tilde{N}) = 0$ for $i > 0$. In fact, for any $\mathcal{O}_X$-module F, $\underline{\mathrm{Ext}}^i_{\mathcal{O}_X}(\tilde{A}, F) = 0$, since $\tilde{A} = \mathcal{O}_X$, and $\underline{\mathrm{Hom}}_{\mathcal{O}_X}(\mathcal{O}_X, F) = F$ is an exact functor in F. (Recall that the $\underline{\mathrm{Ext}}^i_{\mathcal{O}_X}$ are defined as derived functors in the category of $\mathcal{O}_X$-modules, with respect to the second variable.)

Proof of Theorem 2.8 In the first assertion, the question is local, so we may assume that X is the spectrum of a Noetherian ring A. Let $\mathcal{J} = \tilde{I}$ and $F = \tilde{N}$. Then we must show that the homomorphisms

$$\varinjlim_n \underline{\mathrm{Ext}}^i_{\mathcal{O}_X}(\widetilde{A/I^n}, \tilde{N}) \to \underline{H}^i_Y(\tilde{N})$$

are isomorphisms. By proposition 2.2, the right-hand side is equal to $H^i_Y(X,\tilde{N})^\sim$. Moreover, using Lemma 2.12 and noting that direct limits commute with the operation $\sim$ [5, I, 1.3.9 (iii)], we reduce to showing that the homomorphisms

$$\varinjlim_n \mathrm{Ext}^i_A(A/I^n, N) \to H^i_Y(X, \tilde{N})$$

are isomorphisms. This is a homomorphism of cohomological functors from the category of A-modules into itself. For $i = 0$ the functors are isomorphic, their common value being the set of elements of N annihilated by some power of I. For $i > 0$ and N injective, both vanish: The one on the left since the Ext's are derived functors; the one on the right by Corollary 2.7 and Proposition 1.10. This characterizes both functors as derived functors, so they are isomorphic.

For the second assertion, we express $\mathrm{Ext}^i_{\mathcal{O}_X}(\mathcal{O}_n, F)$ and $H^i_Y(X, F)$ as the abutments of spectral sequences, using propositions 1.4 and 2.10. Now one observes that a direct limit of spectral sequences is a spectral sequence, and that we have functorial homomorphisms of spectral sequences, which give the following commutative diagram:

$$\begin{array}{ccc} E_2^{pq} = \varinjlim_n H^p(X, \underline{\mathrm{Ext}}^q_{\mathcal{O}_X}(\mathcal{O}_n, F)) & \to & H^p(X, \underline{H}^q_Y(F)) = E_2'^{pq} \\ \Downarrow & & \Downarrow \\ \varinjlim_n \mathrm{Ext}^i_{\mathcal{O}_X}(\mathcal{O}_n, F) & \to & H^i_Y(X, F) \end{array} \qquad (*)$$

To show that the homomorphism of the abutments is an isomorphism, it will be sufficient to show that the homomorphisms of the E_2^{pq} terms are isomorphisms. But this follows from the isomorphism $(\underline{*})$ established above, and from Proposition 2.11, since X is assumed Noetherian.

§ 3. Relation to Depth

In this section, we recall the notion of depth or homological codimension introduced by Auslander and Buchsbaum [1]. Then we give a theorem (Theorem 3.8) relating the notion of depth to the local cohomology groups.

First let us recall some notations and definitions. If A is a commutative ring, and I is an ideal of A, we denote by V(I) the variety of I, which is the set of all prime ideals of A containing I. V(I) is a closed subset of Spec A.

If N is an A-module, we denote by Supp N the support of N, which is the set of prime ideals $\mathfrak{p}$ of A such that $N_{\mathfrak{p}} \neq 0$. We denote by Ass N the set of associated primes of N: They are those prime ideals $\mathfrak{p}$ of A such that N contains a submodule N_1 isomorphic to $A/\mathfrak{p}$.

If A is a commutative ring, and M an A-module, then a sequence of elements $f_1, \ldots, f_n$ of A is said to be M-regular if for each $i = 1, \ldots, n$, f_i is not a zero-divisor in the module $M/(f_1, \ldots, f_{i-1})M$. (In particular, this means f_1 is not a zero-divisor in M.)

Lemma 3.1. Let A be a Noetherian ring, let I be an ideal, and let M be an A-module of finite type. Then the following conditions are equivalent:

(i) $\mathrm{Hom}(N, M) = 0$ for all A-modules of finite type N whose support is contained in $V(I)$.

(ibis) $\mathrm{Hom}(N, M) = 0$ for some A-module of finite type N whose support is equal to $V(I)$.

(ii) No associated prime of M contains I.

(iii) $\exists f \in I$ such that f is M-regular.

<u>Sublemma 3.2</u>. Let N, M be modules of finite type over a Noetherian ring A. Then

$$\mathrm{Ass}\,(\mathrm{Hom}(N, M)) = \mathrm{Supp}\, N \cap \mathrm{Ass}\, M \quad .$$

<u>Proof</u>. Using well-known facts about associated primes of modules of finite type over Noetherian rings, we have

$$\mathrm{Ass}(\mathrm{Hom}(N, M)) \subseteq \mathrm{Supp}\,(\mathrm{Hom}(N, M)) \subseteq \mathrm{Supp}\, N \quad .$$

Moreover, if we express N as a quotient of a free module

$$A^r \to N \to 0 \quad ,$$

then we have

$$0 \to \mathrm{Hom}(N, M) \to \mathrm{Hom}(A^r, M) = M^r \quad ,$$

from which follows

$$\mathrm{Ass}(\mathrm{Hom}(N, M)) \subseteq \mathrm{Ass}\, M^r = \mathrm{Ass}\, M \quad .$$

Thus we have an inclusion

$$\mathrm{Ass}\,(\mathrm{Hom}(N, M)) \subseteq \mathrm{Supp}\, N \cap \mathrm{Ass}\, M \quad .$$

Conversely, let $\mathfrak{p}$ be a prime ideal of A such that $\mathfrak{p} \in \text{Supp}\, N \cap \text{Ass}\, M$. Then since $\mathfrak{p} \in \text{Ass}\, M$, M has a submodule $M_1 \cong A/\mathfrak{p}$. Since $\mathfrak{p} \in \text{Supp}\, N$, the support of $N/\mathfrak{p}N$ must be all of Spec $A/\mathfrak{p}$. In particular (since N is of finite type), $N/\mathfrak{p}N$ has positive rank over the integral domain $A/\mathfrak{p}$. Hence it has a quotient module N' which is torsion-free and of rank one over $A/\mathfrak{p}$, and therefore isomorphic to an ideal of $A/\mathfrak{p}$. Thus there is an injection $N' \to M_1$, and we can define $\phi \in \text{Hom}(N, M)$ to be the composition of the following homomorphisms

$$N \to N/\mathfrak{p}N \to N' \to M_1 \to M \quad .$$

By our construction, the submodule of Hom(N, M) generated by ϕ is isomorphic to $A/\mathfrak{p}$, so $\mathfrak{p} \in \text{Ass}\,(\text{Hom}\,(N, M))$. Q.E.D.

Proof of Lemma 3.1.

(i) => (ibis). Trivial.

(ibis) => (ii). If Hom(N, M) = 0, and Supp N = V(I), then by the sublemma, $V(I) \cap \text{Ass}\, M = \text{Ass}\,(\text{Hom}\,(N, M)) = \emptyset$, so no associated prime of M contains I.

(ii) => (i). If $\text{Supp}\, N \subseteq V(I)$, then again by the sublemma, $\text{Ass}(\text{Hom}(N, M)) \subseteq V(I) \cap \text{Ass}\, M = \emptyset$ by hypothesis, so Hom (N, M) = 0.

(ii) <=> (iii). Let $\mathfrak{p}_1, \ldots, \mathfrak{p}_r$ be the associated primes of M (which are finite in number since M is of finite type). Then to say none of the $\mathfrak{p}_i$ contains I is the same as to say their union $\bigcup_{i=1}^{r} \mathfrak{p}_i$ does not contain I, which is the same as to say that there is an element $f \in I$ which is in none of the $\mathfrak{p}_i$. But to say that an element $f \in A$ is M-regular is the same as to say that it is in none of the $\mathfrak{p}_i$, so we are done.

<u>Proposition 3.3</u>. Let A be a Noetherian ring, let I be an ideal of A, let M be an A-module of finite type, and let n be an integer. Then the following conditions are equivalent:

(i) $\mathrm{Ext}^i_A(N, M) = 0$ for all A-modules N of finite type such that $\mathrm{Supp}\, N \subseteq V(I)$, and for all integers $i < n$.

(i bis) $\mathrm{Ext}^i_A(N, M) = 0$ for some A-module N of finite type such that $\mathrm{Supp}\, N = V(I)$, and for all integers $i < n$.

(ii) $\exists$ elements $f_1, \ldots, f_n \in I$ forming an M-regular sequence.

<u>Proof</u>. (i) => (ibis). Trivial.

(ibis) => (ii). We proceed by induction on n. If $n \leq 0$, there is nothing to prove. So suppose $n > 0$. Then in particular $\mathrm{Hom}(N, M) = 0$, so by the lemma there exists an $f \in I$ which is M-regular. This f gives rise to an exact sequence as follows:

$$0 \to M \xrightarrow{f} M \to M/fM \to 0 \quad ,$$

where the first map is multiplication by f. Now from our hypothesis and from the exact sequence of Ext's, it follows that $\mathrm{Ext}^i_A(N, M/fM) = 0$ for $i < n - 1$. Therefore, by the induction hypothesis, there are elements $f_2, \ldots, f_n \in I$ which form an M/fM-regular sequence. Then clearly $f, f_2, \ldots, f_n$ is an M-regular sequence of n elements of I.

(ii) => (i). We proceed by induction on n. If $n \leq 0$, there is nothing to prove. So suppose $n > 0$. Then in particular f_1 is M-regular, so there is an exact sequence.

$$0 \to M \xrightarrow{f_1} M \to M/f_1M \to 0 \quad .$$

Now $f_2, \ldots, f_n$ is an M/f_1M-regular sequence of $n - 1$ elements of I, so by the induction hypothesis it follows that $\mathrm{Ext}^i_A(N, M/f_1M) = 0$ for all A-modules of finite type N with support in $V(I)$, and for all integers $i < n - 1$. By the exact sequence of Ext's, this implies that the natural map

$$\mathrm{Ext}^i_A(N, M) \xrightarrow{f_1} \mathrm{Ext}^i_A(N, M)$$

is injective for all $i < n$. But since $f_1 \in I$, f_1 kills N, so this map is also the zero map. Hence $\mathrm{Ext}^i_A(N, M) = 0$ for all $i < n$.

Remark. Note that the lemma (with its Noetherian hypotheses) is used only in the implication (ibis) => (ii) of the proof. The implication (ii) => (i) is valid without assuming A Noetherian nor N, M of finite type. Moreover, one may replace the hypothesis "Supp N $\subseteq$ V(I)" by "$f_1, \dots, f_n$ are nilpotent in N".

Definition. Let A be a Noetherian ring, let I be an ideal of A, and let M be an A-module of finite type. Then the I-depth of M, $\text{depth}_I M$, is the largest integer n such that there exist elements $f_1, \dots, f_n \in I$ which form an M-regular sequence. (Note that $n \leq \dim M$, and is therefore finite.) If A is a local ring, and I is the maximal ideal, we say simply depth M. (The I-depth of M is what Auslander and Buchsbaum [1] call the I-codimension of M.)

Corollary 3.4. Let A, I, M be as above. Then all maximal M-regular sequences of elements of I have the same number of elements, namely, $\text{depth}_I M$.

Corollary 3.5. Let A, I, M be as above. If $f \in I$ is M-regular, then

$$\text{depth}_I M = \text{depth}_I M/fM + 1 \quad .$$

Corollary 3.6. Let A, I, M be as above. Then

$$\operatorname{depth}_I M = \inf_{\mathfrak{p} \in V(I)} (\operatorname{depth} M_{\mathfrak{p}})$$

where $M_{\mathfrak{p}}$ is considered as a module over the local ring $A_{\mathfrak{p}}$, and its depth is the (usual) depth with respect to the maximal ideal $\mathfrak{p}A_{\mathfrak{p}}$.

Proof. If $f_1, \ldots, f_n \in I$ form an M-regular sequence, then the canonical images $\bar{f}_1, \ldots, \bar{f}_n$ of the f_i in $A_{\mathfrak{p}}$ are in $\mathfrak{p}A_{\mathfrak{p}}$, and form an $M_{\mathfrak{p}}$-regular sequence for any $\mathfrak{p} \in V(I)$. Therefore

$$\operatorname{depth}_I M \leq \operatorname{depth} M_{\mathfrak{p}}$$

for each $\mathfrak{p} \in V(I)$.

For the converse, we will prove the following statement: "If n' is an integer such that for each $\mathfrak{p} \in V(I)$, $n' \leq \operatorname{depth} M_{\mathfrak{p}}$, then $n' \leq \operatorname{depth}_I M$". We proceed by induction on n', the case $n' \leq 0$ being trivial. So suppose $n' > 0$. Then for each $\mathfrak{p} \in V(I)$, depth $M_{\mathfrak{p}} \geq 1$, so $\mathfrak{p}A_{\mathfrak{p}}$ is not associated to $M_{\mathfrak{p}}$, i.e., $\mathfrak{p}$ is not associated to M. Then by Lemma 3.1, there exists an $f \in I$ which is M-regular. Now using Corollary 3.5 and the induction hypothesis, we find

$$n' - 1 \leq \operatorname{depth} (M/fM)$$

for each $\mathfrak{p} \in V(I)$. Hence

$$n' - 1 \leq \text{depth}_I M/fM \quad ,$$

which implies that $n' \leq \text{depth}_I M$.

Definition. Let X be a locally Noetherian prescheme, Y a closed subset of X, and F a coherent sheaf on X. Then the Y-depth of F, or $\text{depth}_Y F$, is the $\inf_{x \in Y} (\text{depth } F_x)$.

Remark. If X is an affine scheme, say $X = \text{Spec } A$, $Y = V(I)$ and $F = \tilde{M}$, then $\text{depth}_Y F = \text{depth}_I M$, by Corollary 3.6. We observe thus that the notion of I-depth depends only on the radical of the ideal I.

Restating the above theory in the language of preschemes, we have

Proposition 3.7. Let X be a locally Noetherian prescheme, let Y be a closed subset, let F be a coherent sheaf on X and let n be an integer. Then the following conditions are equivalent:

(i) $\underline{\text{Ext}}^i_{\mathcal{O}_X}(G, F) = 0$ for all coherent sheaves G with $\text{Supp } G \subseteq Y$, and for all integers $i < n$.

(ibis) $\underline{\text{Ext}}^i_{\mathcal{O}_X}(G, F) = 0$ for some coherent sheaf G with $\text{Supp } G = Y$, and for all integers $i < n$.

(ii) $\operatorname{depth}_Y F \geq n$.

(iibis) depth $F_x \geq n$ for all $x \in Y$.

Now we have a theorem which relates the notion of depth to the local cohomology sheaves:

<u>Theorem 3.8</u>. Let X be a locally Noetherian prescheme, let Y be a closed subset, let F be a coherent sheaf on X, and let n be an integer. Then the following conditions are equivalent:

(i) $\underline{H}^i_Y(F) = 0$ for all $i < n$;

(ii) $\operatorname{depth}_Y F \geq n$.

<u>Proof</u>. (i) => (ii). We proceed by induction on n. If $n \leq 0$, there is nothing to prove. So suppose $n > 0$. Then, in particular, condition (i) is satisfied for $n - 1$, so by the induction hypothesis, $\operatorname{depth}_Y F \geq n - 1$. Therefore, by proposition 3.7, $\underline{\operatorname{Ext}}^i_{\mathcal{O}_X}(G, F) = 0$ for all G in the category $\mathcal{C}_Y$ of coherent sheaves on X with support in Y, and for all $i < n - 1$. In other words, the functor

$$G \rightsquigarrow \underline{\operatorname{Ext}}^{n-1}_{\mathcal{O}_X}(G, F)$$

from $\mathcal{C}_Y$ to sheaves on X is left-exact. Let $\mathcal{J}_Y$ be the sheaf of ideals of Y, and for each positive integer m, let $\mathcal{O}_m = \mathcal{O}_X/\mathcal{J}_Y^m$. Then the sheaves $\mathcal{O}_m$ are in the category $\mathcal{C}_Y$, so the surjections

$$\mathcal{O}_{m'} \to \mathcal{O}_m \to 0$$

for $m' \geq m$ give rise to injections

$$0 \to \underline{\mathrm{Ext}}^{n-1}_{\mathcal{O}_X}(\mathcal{O}_m, F) \to \underline{\mathrm{Ext}}^{n-1}_{\mathcal{O}_X}(\mathcal{O}_{m'}, F) .$$

In other words, all the homomorphisms in the direct system $(\underline{\mathrm{Ext}}^{n-1}_{\mathcal{O}_X}(\mathcal{O}_m, F))_m$ are injections. By Theorem 2.8, the direct limit of this system is $\underline{H}_Y^{n-1}(F)$, which is zero by our hypothesis. Hence each of the terms of our direct system is zero, in particular, $\underline{\mathrm{Ext}}^{n-1}_{\mathcal{O}_X}(\mathcal{O}/\mathcal{J}_Y, F) = 0$. We have already seen that $\underline{\mathrm{Ext}}^{i}_{\mathcal{O}_X}(\mathcal{O}/\mathcal{J}_Y, F) = 0$ for $i < n-1$, so from Proposition 3.7 we conclude that $\mathrm{depth}_Y F \geq n$.

(ii) => (i). Suppose that $\mathrm{depth}_Y F \geq n$. Using the above notation, by Theorem 2.8 we have for any i,

$$\underline{H}_Y^i(F) \cong \varinjlim_m \underline{\mathrm{Ext}}^i_{\mathcal{O}_X}(\mathcal{O}_m, F) .$$

By Proposition 3.7, these Ext's are 0 for $i < n$, since the sheaves $\mathcal{O}_m$ have support in Y, hence $\underline{H}_Y^i(F) = 0$ for $i < n$. Q.E.D.

Corollary 3.9. Let X be a connected locally Noetherian prescheme, and Y a closed subprescheme such that $\operatorname{depth}_Y \mathcal{O}_X \geq 2$. Then X - Y is connected.

Proof. By the theorem, our hypothesis implies that $\underline{H}^i_Y(\mathcal{O}_X) = 0$ for $i = 0, 1$. By Proposition 1.11, this implies that the map

$$H^0(X, \mathcal{O}_X) \to H^0(X - Y, \mathcal{O}_X)$$

is bijective. To complete the proof of the corollary, we need only observe that a prescheme X is connected if and only if $H^0(X, \mathcal{O}_X)$, considered as a module over itself, is indecomposable. For then X connected => $H^0(X, \mathcal{O}_X) = H^0(X - Y, \mathcal{O}_X)$ is indecomposable => X - Y is connected.

If X is disconnected, $H^0(X, \mathcal{O}_X)$ is the direct sum of $H^0(U_i, \mathcal{O}_X)$, where U_i are the components, so $H^0(X, \mathcal{O}_X)$ is decomposable. Conversely, suppose $H^0(X, \mathcal{O}_X)$ is decomposable. Then there are nontrivial idempotents $e_1, e_2 \in H^0(X, \mathcal{O}_X)$, that is, elements e_1, e_2 different from zero, and such that $1 = e_1 + e_2$; $e_1 e_2 = 0$; $e_1^2 = e_1$, $e_2^2 = e_2$. For each $x \in X$, the local ring $\mathcal{O}_x$ is indecomposable, hence one of the e_i is zero at x. Neither can be zero at every point of X, hence X is the disjoint sum of the closed sets where e_1 is zero and where e_2 is zero. Thus X is disconnected.

Remark. The Corollary applies in particular if Y is of codimension at least two and $\mathcal{O}_X$ has the property S_2 of Serre, as it does for example if X is normal, or if X is a complete intersection in a non-singular ambient prescheme. For details, see Hartshorne [8], who proves the corollary directly from the definition of depth, and applies it to the study of complete intersections.

In the particular case of a local Noetherian ring, the above results give the following

Corollary 3.10. Let A be a local Noetherian ring with maximal ideal m and residue field k. Let $X = \operatorname{Spec} A$, $U = X - \{m\}$. Let M be a module of finite type, and let $n \geq 0$ be an integer. Then the following conditions are equivalent

(i) depth $M > n$.

(ii) $\operatorname{Ext}^i(k, M) = 0$ for $i \leq n$.

(iii) $H^i(X, \tilde{M}) \to H^i(U, \tilde{M})$ is bijective for $i < n$ and injective for $i = 0$.

(iv) $H^i_{\{m\}}(\tilde{M}) = 0$ for $i \leq n$.

§4. Functors on A-modules

In this section, A will denote a commutative Noetherian ring. We will discuss various left-exact functors from the category of A-modules to the category of abelian groups. In particular, we will discuss dualizing functors which will be used later in the section on duality.

As a matter of notation, let

(Ab) = the category of abelian groups ,

$\mathcal{C}$ = the category of A-modules ,

$\mathcal{C}^f$ = the category of A-modules of finite type .

If $\mathfrak{m}$ is an ideal of A, let

$\mathcal{C}_{\mathfrak{m}}$ = the category of A-modules with support in $V(\mathfrak{m})$,

$\mathcal{C}^f_{\mathfrak{m}}$ = the category of A-modules of finite type with support in $V(\mathfrak{m})$.

If $\mathfrak{a}$ is any category, we will denote by $\mathfrak{a}^\circ$ the opposed category.

<u>Lemma 4.1.</u> Let T be a contravariant additive functor from $\mathcal{C}$ to (Ab). Then there is a natural functorial morphism

$$\phi : T \to \mathrm{Hom}_A(\cdot\,,\ T(A)) .$$

Proof. For any $m \in M$, let $\epsilon_m : A \to M$ be the morphism which sends 1 into m. Then $T(\epsilon_m)$ is a morphism of $T(M) \to T(A)$. For a fixed element of $T(M)$, as $m \in M$ varies, we have an element of $\mathrm{Hom}_A(M, T(A))$. Thus we have defined a morphism.

$$\phi(M) : T(M) \to \mathrm{Hom}_A(M, T(A)) \quad .$$

One checks that it is functorial in M. In fact, if one gives $T(M)$ and $\mathrm{Hom}_A(M, T(A))$ the natural A-module structures, it is a morphism of A-modules.

Proposition 4.2. Let T be a contravariant additive functor from $\mathcal{C}^f$ to (Ab). Then the morphism

$$\phi : T \to \mathrm{Hom}_A(\cdot \ , T(A))$$

is an isomorphism if and only if T is left exact.

Proof. If ϕ is an isomorphism, then T is left-exact since the functor Hom is. On the other hand, suppose T is left-exact. Then in the first place

$$T(A^r) \to \mathrm{Hom}_A(A^r, T(A)) = T(A)^r$$

is an isomorphism for any positive integer r, since T is additive. Now if M is any A-module of finite type, there is an exact sequence

$$A^r \to A^s \to M \to 0 \quad .$$

Letting T' stand for the functor $\mathrm{Hom}_A(\cdot, T(A))$, we have an exact commutative diagram

$$\begin{array}{ccccccc} 0 & \to & T(M) & \to & T(A^s) & \to & T(A^r) \\ & & \downarrow & & \downarrow\approx & & \downarrow\approx \\ 0 & \to & T'(M) & \to & T'(A^s) & \to & T'(A^r) \end{array}$$

whence by the 5-lemma, $T(M) \to T'(M)$ is an isomorphism.

Definition. If $\mathcal{A}$, $\mathcal{B}$ are two abelian categories, we denote by Sex($\mathcal{A}$, $\mathcal{B}$) the category of left-exact (covariant) functors from $\mathcal{A}$ to $\mathcal{B}$. (This notation is due to Gabriel, and is explained thus: s = sinister, ex = exact.)

Corollary 4.3 The categories Sex($\mathcal{C}^{f\circ}$, Ab) and $\mathcal{C}$ are made equivalent by the functors

$$T \rightsquigarrow T(A) \qquad \text{for } T \in \mathrm{Sex}(\mathcal{C}^{f\circ}, \mathrm{Ab}) \quad ,$$

and

$$I \rightsquigarrow \mathrm{Hom}_A(\cdot, I) \qquad \text{for } I \in \mathcal{C} \quad .$$

<u>Remark</u>. Note that in the equivalence of Corollary 4.3, the exact functors correspond to the injective modules. Indeed, if I is injective, $\mathrm{Hom}_A(\cdot, I)$ is exact. On the other hand, if T is exact, then $T(A)$ is injective, since over a <u>Noetherian</u> ring, a module I is injective if and only if the functor $M \rightsquigarrow \mathrm{Hom}_A(M, I)$ is exact for all A-modules M <u>of finite type</u>.

<u>Lemma 4.4</u>. Let $\mathfrak{m}$ be an ideal in A, and let T be an additive contravariant functor from $\mathcal{C}_{\mathfrak{m}}^f$ to (Ab). Then there is a natural functorial morphism

$$\phi_{\mathfrak{m}} : T \to \mathrm{Hom}_A(\cdot, I) \quad ,$$

where $I = \varinjlim_n T(A/\mathfrak{m}^n)$.

<u>Proof</u>. Let $M \in \mathcal{C}_{\mathfrak{m}}^f$. Then M is annihilated by some power of $\mathfrak{m}$, say $\mathfrak{m}^n$. In other words, M is a module of finite type over $A/\mathfrak{m}^n$. So by Lemma 4.1 there is a morphism

$$T(M) \to \mathrm{Hom}_{A/\mathfrak{m}^n}(M, T(A/\mathfrak{m}^n)) = \mathrm{Hom}_A(M, T(A/\mathfrak{m}^n)) \quad .$$

Furthermore, there is a natural morphism

$$\mathrm{Hom}_A(M, T(A/\mathfrak{m}^n)) \to \varinjlim_n \mathrm{Hom}_A(M, T(A/\mathfrak{m}^n)) \quad ,$$

and this limit is equal to $\mathrm{Hom}_A(M,I)$, since M is of finite type, and therefore $\mathrm{Hom}_A(M,\cdot)$ commutes with direct limits. Composing these morphisms, one obtains a morphism

$$T(M) \to \mathrm{Hom}_A(M,I) \quad ,$$

which is easily seen to be independent of the n chosen above, and functorial in M.

<u>Proposition 4.5</u>. The morphism $\phi_{\mathfrak{m}}$ of Lemma 4.4 is an isomorphism if and only if T is left exact.

<u>Proof</u>. If $\phi_{\mathfrak{m}}$ is an isomorphism, then T is left exact, since the functor Hom is. On the other hand, suppose T is left exact, and let $M \in \mathcal{C}^f_{\mathfrak{m}}$ be a module annihilated by $\mathfrak{m}^n$. Then for each $n' \geq n$, the morphism

$$T(M) \to \mathrm{Hom}_A(M, T(A/\mathfrak{m}^{n'}))$$

is an isomorphism, by Proposition 4.2. Taking the limit as n' goes to infinity, we find that $\phi_{\mathfrak{m}}$ is also an isomorphism.

<u>Corollary 4.6</u>. The categories $\mathrm{Sex}(\mathcal{C}^{f\,\circ}_{\mathfrak{m}}, \mathrm{Ab})$ and $\mathcal{C}_{\mathfrak{m}}$ are made equivalent by the functors

$$T \rightsquigarrow \varinjlim_n T(A/\mathfrak{m}^n) \qquad \text{for } T \in \mathrm{Sex}(\mathcal{C}^{f\,\circ}_{\mathfrak{m}}, \mathrm{Ab})$$

and

$$I \rightsquigarrow \mathrm{Hom}_A(\cdot, I) \quad \text{for } I \in \mathcal{C}_{\mathfrak{m}} .$$

Proof. We need only check that $\varinjlim_n \mathrm{Hom}_A(A/\mathfrak{m}^n, I) \cong I$. Now $\mathrm{Hom}(A/\mathfrak{m}^n, I)$ is isomorphic to the submodule of I consisting of those elements annihilated by $\mathfrak{m}^n$. I is the union of these submodules since it was assumed to have support in $V(\mathfrak{m})$.

Proposition 4.7. In the equivalence of Corollary 4.6, the exact functors correspond to the injective modules.

Proof. If I is injective, then $\mathrm{Hom}_A(\cdot, I)$ is exact. Conversely, suppose T is exact, and let $I = \varinjlim_n T(A/\mathfrak{m}^n)$. We wish to show I injective. It will be sufficient to prove that if $\mathfrak{a}$ is an ideal in A, and $f : \mathfrak{a} \to I$ a morphism, then f extends to a morphism, $\bar{f} : A \to I$.

So suppose $\mathfrak{a}$ is an ideal of A, and $f: \mathfrak{a} \to I$ a morphism. Since $\mathfrak{a}$ is of finite type (A being Noetherian), and $I \in \mathcal{C}_{\mathfrak{m}}$, f annihilates $\mathfrak{m}^n\mathfrak{a}$ for some n. By Krull's theorem, the $\mathfrak{m}$-topology of $\mathfrak{a}$ is induced by the $\mathfrak{m}$-topology of A, i.e., there is an integer r such that $\mathfrak{m}^r \cap \mathfrak{a} \subseteq \mathfrak{m}^n\mathfrak{a}$. Thus f also annihilates $\mathfrak{m}^r \cap \mathfrak{a}$, and therefore f factors as follows:

$$\mathfrak{a} \xrightarrow{p} \mathfrak{a}/\mathfrak{m}^r \cap \mathfrak{a} \xrightarrow{f_1} I .$$

But now we have an injection

$$0 \to \mathfrak{a}/\mathfrak{m}^r \cap \mathfrak{a} \to A/\mathfrak{m}^r$$

of modules in $\mathcal{C}^f_{\mathfrak{m}}$; we have a morphism

$$f_1 : \mathfrak{a}/\mathfrak{m}^r \cap \mathfrak{a} \to I ;$$

and T is exact. Therefore, f_1 extends to a morphism

$$\overline{f}_1 : A/\mathfrak{m}^r \to I .$$

If $\overline{p} : A \to A/\mathfrak{m}^r$ is the canonical projection morphism, then $\overline{f} = \overline{f}_1 \circ \overline{p}$ is the desired extension of f. Hence I is injective.

Corollary 4.8. Let I be an injective A-module, let $\mathfrak{m}$ be an ideal of A, and let $I_{\mathfrak{m}} = H^{\bullet}_{\mathfrak{m}}(I)$ be the largest submodule of I with support in $V(\mathfrak{m})$. Then $I_{\mathfrak{m}}$ is injective.

Proof. Let $T \varepsilon \operatorname{Sex}(\mathcal{C}^{f\,\circ}_{\mathfrak{m}}, Ab)$ be the exact functor $\operatorname{Hom}_A(\cdot, I)$. Then by Proposition 4.7, $I_{\mathfrak{m}} = \varinjlim T(A/\mathfrak{m}^n)$ is injective.

We now come to the study of dualizing functors.

Proposition 4.9. Let $\mathfrak{m}$ be an ideal of finite colength in A (i.e., such that $A/\mathfrak{m}$ is an Artin ring), and let $T \in \mathrm{Sex}(\mathcal{C}_{\mathfrak{m}}^{f\,\bullet}, \mathrm{Ab})$. Then the following conditions are equivalent:

(i) For all $M \in \mathcal{C}_{\mathfrak{m}}^{f}$, $T(M)$ is an A-module of finite type, and the natural morphism

$$M \to TT(M)$$

(defined via the isomorphism of Lemma 4.4) is an isomorphism.

(ii) T is exact, and for each field k of the form $A/\mathfrak{m}_i$, where $\mathfrak{m}_i$ is a maximal ideal containing $\mathfrak{m}$, there is some isomorphism $T(k) \cong k$.

Proof. (i) => (ii). Let k be a field of the form $A/\mathfrak{m}_i$ as above. Then $T(k)$ is of finite type by hypothesis; moreover, it is annihilated by $\mathfrak{m}_i$, hence must be isomorphic to k^n for some $n \geq 0$. Since T is additive, $TT(k) = k^{n^2}$, whence $n = 1$. Thus $T(k) \cong k$.

It remains to show T is exact. We first show that T is faithful, i.e., if $Q \neq 0$, then $T(Q) \neq 0$. Let $Q \in \mathcal{C}_{\mathfrak{m}}^{f}$ be non-zero, then there is a surjection

$$Q \to k \to 0 \quad ,$$

where k is a field of the form $A/\mathfrak{m}_i$. Since T is exact,

$$0 \to T(k) \to T(Q).$$

But $T(k) \cong k \neq 0$, hence also $T(Q) \neq 0$.

Now T is left exact by hypothesis, so we need only show that it is right exact. Let

$$0 \to M' \to M$$

be an injection of modules in $\mathcal{C}_{\mathfrak{m}}^{f}$. Apply T, and let Q be the cokernel:

$$T(M) \to T(M') \to Q \to 0 \quad .$$

Since T is left exact, we have an exact sequence

$$0 \to T(Q) \to TT(M') \to TT(M) \quad .$$

But TT is isomorphic to the identity functor by hypothesis. Therefore, $T(Q) = 0$, and since T is faithful, $Q = 0$, i.e., T is exact.

(ii) => (i). If $M \in \mathcal{C}_{\mathfrak{m}}^{f}$, then M has a finite composition series whose quotients are all fields k of the form $A/\mathfrak{m}_i$, as above. Therefore, since T is exact, $T(M)$ will have a finite composition series whose quotients are of the form $T(k) \cong k$. Hence $T(M)$ is of finite length (in fact, its length is equal to the length of M).

For the second assertion of (i), we first consider the case where M is a field k of the form $A/\mathfrak{m}_i$. Then $TT(k) \cong k$ in some way, and the natural map $k \to TT(k)$ is easily seen not to be zero, hence it is an isomorphism.

Now if M is any module in $\mathcal{C}^f_{\mathfrak{m}}$, and

$$0 \to M' \to M \to M'' \to 0$$

is an exact sequence, then we deduce an exact sequence

$$0 \to TT(M') \to TT(M) \to TT(M'') \to 0,$$

together with a canonical morphism of the first exact sequence into the second. If the two outside arrows are isomorphism, then by the 5-lemma, so is the middle one. Thus by induction on the length of M (since any $M \in \mathcal{C}^f_{\mathfrak{m}}$ has finite length), we see that the natural morphism

$$M \to TT(M)$$

is an isomorphism for all $M \in \mathcal{C}^f_{\mathfrak{m}}$. Q.E.D.

Definition. Let A be a commutative Noetherian ring, and $\mathfrak{m}$ a maximal ideal. Then a functor $T \in \mathrm{Sex}(\mathcal{C}^{f\,\circ}_{\mathfrak{m}}, \mathrm{Ab})$, satisfying the equivalent conditions of Proposition 4.9, is called a dualizing functor for A at $\mathfrak{m}$. An injective module I with support in

$V(\mathfrak{m})$ is called a <u>dualizing module for</u> A <u>at</u> $\mathfrak{m}$, if the functor $T = \mathrm{Hom}(\cdot, I)$ is a dualizing functor for A at $\mathfrak{m}$. (By Proposition 4.7, there is a natural one-to-one correspondence between these dualizing functors and dualizing modules.) If A is a local ring with maximal ideal $\mathfrak{m}$, we say simply "dualizing functor" or "dualizing module," with $\mathfrak{m}$ understood.

<u>Remarks</u>. 1) Let A be a Noetherian ring, and $\mathfrak{m}$ a maximal ideal. Then a dualizing functor for A at $\mathfrak{m}$ is the same thing as a dualizing functor for the local ring $A_{\mathfrak{m}}$ at its maximal ideal $\mathfrak{m}A_{\mathfrak{m}}$. Indeed, the categories $\mathcal{C}^f_{\mathfrak{m}}$ involved are isomorphic.

2) Similarly, let A be a local ring, and $I \varepsilon \mathcal{C}_{\mathfrak{m}}$ an A-module. Then I has a natural structure of $\hat{A}$-module, and I is dualizing for A if and only if it is dualizing for $\hat{A}$. Again the categories $\mathcal{C}^f_{\mathfrak{m}}$ involved are isomorphic.

3) Let A be a Noetherian ring, $\mathfrak{m}$ a maximal ideal, and I and A-module. For each $n = 1, 2, \ldots$, let I_n be the $A/\mathfrak{m}^n$-module $\mathrm{Hom}_A(A/\mathfrak{m}^n, I)$. Then I is dualizing for A at $\mathfrak{m}$ if and only if for all large enough n (and hence for all n), I_n is dualizing for $A/\mathfrak{m}^n$ at the maximal ideal $\mathfrak{m}/\mathfrak{m}^n$. Indeed,

observe that if $M \in \mathcal{C}^f_{\mathfrak{m}}$ is annihilated by $\mathfrak{m}^n$, then $\mathrm{Hom}_A(M,I) \cong \mathrm{Hom}_{A/\mathfrak{m}^n}(M,I_n)$. Thus $\mathrm{Hom}_A(\cdot,I)$ is exact for all $M \in \mathcal{C}^f_{\mathfrak{m}}$ if and only if for all large enough n, $\mathrm{Hom}_{A/\mathfrak{m}^n}(M, I_n)$ is exact for $A/\mathfrak{m}^n$-modules of finite type. Moreover, $\mathrm{Hom}_A(k,I) \cong \mathrm{Hom}_{A/\mathfrak{m}^n}(k,I_n)$ for any n, where $k = A/\mathfrak{m}$. Thus the result follows from criterion (ii) of Proposition 4.9.

We now recall, for the sake of convenience, the definition of an injective hull.

Definition. An injection $M \to P$ of A-modules is an essential extension if whenever N is a submodule of P such that $M \cap N = 0$, then $N = 0$.

Definition. If M is an A-module, an injective hull of M is an essential extension $M \to I$ where I is injective. By abuse of language, we also call I an injective hull of M.

Theorem. (Eckmann-Schopf [2]; see also [3, Ch. II, p. 20 ff.]) Let A be a ring with identity. Then every A-module has an injective hull, unique to within (non-unique) isomorphism.

Proposition 4.10. Let A be a Noetherian local ring. Then an A-module I is dualizing for A if and only if it is an injective hull of the residue field k.

Proof. First suppose that I is dualizing. Then $\mathrm{Hom}(k, I) \cong k$, and so in particular, we can embed k in I. To show that $k \to I$ is an essential extension, let Q be a submodule of I such that $k \cap Q = 0$. Then since $\mathrm{Hom}(k, I) \cong k$, we must have $\mathrm{Hom}(k, Q) = 0$. But I, hence also Q, has support in $V(\mathfrak{m})$, so $Q = 0$. Now I dualizing implies that I is injective, so I is an injective hull of k.

On the other hand, suppose I is an injective hull of k. Then any homomorphism of k into I must have its image in k, because $k \to I$ is an essential extension, and k is a field. Thus $\mathrm{Hom}(k, I) \cong k$, and I is a dualizing module.

Corollary 4.11. Let A be a Noetherian ring, and $\mathfrak{m}$ a maximal ideal. Then A has a dualizing module at $\mathfrak{m}$, unique to within (non-unique) isomorphism

Example. Let $A = \mathbb{Z}$, and let $\mathfrak{m} = p\mathbb{Z}$, where p is a prime. Then a dualizing module for A at $\mathfrak{m}$ is the group $\mathbb{Z}_{p^\infty}$, whose automorphisms are the units in the ring of p-adic integers.

Let A be a Noetherian local ring, and D a dualizing functor. We list without proof various properties of D, most of which follow from the above discussion. For more details, see Matlis [9].*

1. D gives an isomorphism of the category $\mathcal{C}^f_{\mathfrak{m}}$ with its dual category $\mathcal{C}^{f\circ}_{\mathfrak{m}}$. D is exact and preserves length. It gives a 1-1 correspondence between submodules of M, and quotient modules of D(M), and vice versa, for any $M \in \mathcal{C}^f_{\mathfrak{m}}$. Finally, D^2 is isomorphic to the identity.

2. D transforms direct systems of $\mathcal{C}^f_{\mathfrak{m}}$ into inverse systems, and vice versa. Thus, passing to limits, D gives a correspondence between direct limits and inverse limits of modules of finite length over A.

3. In particular, the functor $D = \mathrm{Hom}(\cdot, I)$, where I is a dualizing module, gives an anti-isomorphism of the categories

(ACC) = the category of $\hat{A}$-modules of finite type, and

(DCC) = the category of A-modules M satisfying the following three equivalent conditions:

(i) M has the descending chain condition,

(ii) M is of <u>co-finite type</u>, i.e., Hom(k, M) is a finite-dimensional vector space over k, and M has support in $V(\mathfrak{m})$.

(iii) M is a submodule of I^n, for some n.

In this correspondence, $D(\hat{A}) = I$; $D(I) = \mathrm{Hom}(I, I) = \hat{A}$. The ideals $\mathfrak{a}$ of $\hat{A}$ correspond to submodules of I by the map

---

*See also Macdonald [21], who extends the range of the dualizing functor to "linearly compact" modules.

$$\mathfrak{a} \rightsquigarrow D(\hat{A}/\mathfrak{a}),$$

and the ideals of definition correspond to the submodules of I of finite length.

Proposition 4.12. Let $B \to A$ be a surjective morphism of local Noetherian rings, and let $\mathfrak{m}$ be the maximal ideal of B. Then

1) If D is a dualizing functor for B, then $\dot{D}$ restricted to A-modules is a dualizing functor for A.

2) If I is a dualizing module for B, then $I' = \mathrm{Hom}_B(A, I)$ is a dualizing module for A.

Proof. The first statement follows immediately from the definition of a dualizing functor (see Proposition 4.9 (ii)). Hence the functor $\mathrm{Hom}_B(\cdot, I)$ is a dualizing functor for A, and its associated dualizing module is $\varinjlim_n \mathrm{Hom}_B(A/(\mathfrak{m}A)^n, I) = I'$, since any morphism of A into I annihilates some power of $\mathfrak{m}A$.

We now give three specific examples of dualizing functors.

Example 1. Let A be a Noetherian local ring containing a field k_0, and suppose that the residue field k is a finite extension of k_0. Then any module $M \varepsilon \mathcal{C}^f_{\mathfrak{m}}$ is a finite-dimensional vector space over k_0, and we can define a dualizing functor D by

$$D(M) = \mathrm{Hom}_{k_0}(M, k_0) \quad .$$

The corresponding dualizing module is

$$I = \varinjlim \mathrm{Hom}_{k_0}(A/\mathfrak{m}^n, k_0) \quad ,$$

which can be thought of as the module of continuous homomorphisms of A with the $\mathfrak{m}$-adic topology into k_0.

Example 2. The case of a Gorenstein ring.*

Definition. A Noetherian local ring A of dimension n, with residue field k, is called a Gorenstein ring if

$$\mathrm{Ext}^i_A(k, A) = \begin{cases} 0 & \text{for } i \neq n \\ k & \text{for } i = n \end{cases} \quad .$$

*See also the paper of Bass [13], and [17, ch. V § 9] for more information about Gorenstein rings.

<u>Examples</u>. Any regular local ring is Gorenstein. Any Gorenstein ring is Cohen-Macaulay. (We recall that a Noetherian local ring A is called a <u>Cohen-Macaulay ring</u> if its depth equals its dimension: See Cor. 3.10.)

<u>Proposition 4.13</u>. Let A be a Gorenstein ring of dimension n. Then the functor

$$M \rightsquigarrow \operatorname{Ext}^n(M,A)$$

is dualizing, and its associated dualizing module is $H^n_{\mathfrak{m}}(A) = H^n_Y(X, \mathcal{O}_X)$, where $X = \operatorname{Spec} A$, and $Y = \{\mathfrak{m}\}$ is the closed point.

<u>Proof</u>. First we show that for any $M \in \mathcal{C}^f_{\mathfrak{m}}$, and for any $i \neq n$, $\operatorname{Ext}^i(M,A) = 0$. Indeed, if

$$0 \to M' \to M \to M'' \to 0$$

is an exact sequence in $\mathcal{C}^f_{\mathfrak{m}}$, then there is an exact sequence

$$\operatorname{Ext}^i(M'',A) \to \operatorname{Ext}^i(M,A) \to \operatorname{Ext}^i(M',A) \quad ,$$

for any i. If the two outside groups are zero, so is the middle one. Thus by induction on the length of M, one shows that

$$\operatorname{Ext}^i(M,A) = 0$$

for all $M \in \mathcal{C}^f_{\mathfrak{m}}$, and all $i \neq n$, since by hypothesis it is true for $M = k$.

It follows now that the functor

$$M \rightsquigarrow \mathrm{Ext}^n(M,A)$$

is exact for $M \in \mathcal{C}^f_{\mathfrak{m}}$. On the other hand, $\mathrm{Ext}^n(k,A) \cong k$ by hypothesis, so we have a dualizing functor. Its dualizing module, by Theorem 2.8, is

$$I = \varinjlim_k \mathrm{Ext}^n(A/\mathfrak{m}^k, A) = H^n_{\mathfrak{m}}(A) \quad .$$

In particular, we have shown that $H^n_{\mathfrak{m}}(A)$ is an injective hull of k.

Proposition 4.14. Let A be a Noetherian local ring of dimension n. Then the following conditions are equivalent:

(i) A is Gorenstein.

(ii) A is Cohen-Macaulay, and $H^n_{\mathfrak{m}}(A)$ is dualizing.

Proof. (i) => (ii) by the previous proposition.

(ii) => (i). Since A is Cohen-Macaulay, $\mathrm{Ext}^i(k,A) = 0$ for $i < n$, by Corollary 3.10. Therefore, we can show as in the proof of the previous proposition that $\mathrm{Ext}^i(M,A) = 0$ for all $M \in \mathcal{C}^f_{\mathfrak{m}}$, and all $i < n$. Therefore, the functor

$$T : M \rightsquigarrow \mathrm{Ext}^n(M,A)$$

is left exact, and so by Proposition 4.5 and Theorem 2.8,

$$\mathrm{Ext}^n(M,A) \cong \mathrm{Hom}\,(M, H^n_{\mathfrak{m}}(A)) \quad .$$

But $H^n_{\mathfrak{m}}(A)$ was assumed to be dualizing. Therefore, T is an exact functor, and $T(k) \cong k$.

It remains to show that $\mathrm{Ext}^i(k,A) = 0$ for $i > n$. Since T is exact, the functor

$$M \rightsquigarrow \mathrm{Ext}^{n+1}(M,A)$$

is left exact for $M \in \mathcal{C}^f_{\mathfrak{m}}$, and so, as above,

$$\mathrm{Ext}^{n+1}(M,A) \cong \mathrm{Hom}\,(M, H^{n+1}_{\mathfrak{m}}(A)) \quad .$$

But $H^{n+1}_{\mathfrak{m}}(A) = 0$ by Proposition 1.12, since $n = \dim\,(\mathrm{Spec}\,A)$. Thus $\mathrm{Ext}^{n+1}(M,A) = 0$ for all $M \in \mathcal{C}^f_{\mathfrak{m}}$, and therefore the functor

$$M \rightsquigarrow \mathrm{Ext}^{n+2}(M,A)$$

is left exact. Proceeding thus by induction on $i > n$, one shows that $\mathrm{Ext}^i(M,A) = 0$ for all $i > n$ and all $M \in \mathcal{C}^f_{\mathfrak{m}}$. Thus A is Gorenstein. Q.E.D.

<u>Exercises</u>. 1) Let A be a Cohen-Macaulay local ring of dimension n, and let $\mathfrak{q}$ be an ideal genera ted by a maximal

A-sequence. Show that $H^n_{\mathfrak{m}}(A)$ is an essential extension of $A/\mathfrak{q}$, and therefore that the dimension e of the vector space $\mathrm{Hom}(k, A/\mathfrak{q})$ is independent of $\mathfrak{q}$. Show also that $e = 1$ for a Gorenstein ring.

2) Let A be as in the previous exercise. Show that $H^n_{\mathfrak{m}}(A)$ is an indecomposable module, and deduce that in Proposition 4.14(ii), it is sufficient to assume $H^n_{\mathfrak{m}}(A)$ injective.

3) Show that if A is a <u>complete intersection</u> (i.e., a quotient of a regular local ring B by an ideal $\mathfrak{b}$ which can be generated by a B-sequence), then A is a Gorenstein ring.

<u>Example 3</u>. Let $A = k[[x_1, \ldots, x_n]]$ be the ring of formal power series in n variables. This falls under the situation of both of the above examples, so we have at hand two dualizing modules, namely $\mathrm{Hom\,cont}_k(A, k)$ and $H^n_{\mathfrak{m}}(A)$. So by the uniqueness of the injective hull, we know there is some non-canonical isomorphism connecting them.

This fact can be interpreted in terms of the theory of differentials. Let Ω^1 be the module of continuous differentials of A/k , which in this case will be the free A-module generated by $dx_1, \ldots, dx_n$. Let $\Omega^n = \Lambda^n(\Omega^1)$ by the nth exterior power of Ω^1, which is called the module of (continuous) n-differentials. Then Ω^n is the free A-module generated by $dx_1 \wedge \ldots \wedge dx_n$. So $H^n_{\mathfrak{m}}(\Omega^n)$ is a dualizing module. Now the residue map*is a continuous homomorphism

$$\text{Res}: H^n_{\mathfrak{m}}(\Omega^n) \to k \quad ,$$

canonically defined, which gives rise to a canonical isomorphism

$$H^n_{\mathfrak{m}}(\Omega^n) \xrightarrow{\sim} \text{Hom cont}_k(A, k)$$

between the two dualizing modules.

*The definition of the residue map is rather subtle. In the terminology of [17, ch. VI, § 4], it is the trace map, applied to the last term of a residual complex, which can be identified with $H^n_{\mathfrak{m}}(\Omega^n)$.

§ 5. Some Applications

In this section we give some applications of the theory developed so far. In particular, we study "the first non-vanishing Ext group"; various properties of base change; and the coherence of direct image sheaves. Some of this material, especially the part on base changes, will be used in Section 6.

Proposition 5.1. Let X be a locally Noetherian prescheme, let Y be a closed subset, let F be a coherent sheaf on X such that $\operatorname{depth}_Y F \geq n$, and let $\mathcal{C}_Y^f$ be the category of coherent sheaves on X with support in Y. Then there is a functorial isomorphism, for all $G \in \mathcal{C}_Y^f$,

$$\underline{\mathrm{Ext}}^n_{\mathcal{O}_X}(G, F) \simeq \underline{\mathrm{Hom}}_{\mathcal{O}_X}(G, \underline{H}^n_Y(F)) .$$

Proof. First note that since $\operatorname{depth}_Y F \geq n$, the sheaves $\underline{\mathrm{Ext}}^i_{\mathcal{O}_X}(G, F) = 0$ for all $i < n$, and for all $G \in \mathcal{C}_Y^f$, by Proposition 3.7. Thus n is the first integer for which $\underline{\mathrm{Ext}}^n_{\mathcal{O}_X}(G, F)$ may not vanish. To establish the sheaf isomorphism of the proposition, we make a definition for the corresponding modules in case X is affine, and show that this definition commutes with localization to a smaller affine. We then invoke proposition 2.2

and Lemma 2.12 to pass from modules to sheaves, and glue together the isomorphisms we have defined over the affine pieces of X.

So suppose that X is the prime spectrum of a Noetherian ring A, and let $Y = V(J)$, $F = \tilde{N}$, where J is an ideal in A, and N an A-module whose J-depth is $\geq n$. Then we have seen that $\mathrm{Ext}^i_A(M,N) = 0$ for $i < n$ and for $M \in \mathcal{C}^f_J$. Thus the functor

$$M \rightsquigarrow \mathrm{Ext}^n_A(M,N)$$

is left exact, and so by Proposition 4.5 there is a canonical functorial isomorphism

$$\mathrm{Ext}^n_A(M,N) \xrightarrow{\sim} \mathrm{Hom}_A(M,I) \quad ,$$

where

$$I = \varinjlim_k \mathrm{Ext}^n_A(A/J^k, N) = H^n_J(N) \quad .$$

Now if S is a multiplicative system in A, and if $B = A_S$, then the isomorphism obtained as above for B-modules is the localization of this one, since localization commutes with Ext's and direct limits.

<u>Proposition 5.2</u>. Let X be a locally Noetherian prescheme, let $n \geq 0$ be an integer, and let G, F be coherent sheaves on X such that $\underline{\mathrm{Ext}}^i_{\mathcal{O}_X}(G,F) = 0$ for $i < n$. Let Z be the set of points $z \in \mathrm{Supp}\ G$ such that depth $F_z = n$. Assume that Z

has no embedded points. Then

$$\mathrm{Ass}\ \underline{\mathrm{Ext}}^n_{\mathcal{O}_X}(G, F) = Z \quad .$$

Lemma 5.3. Let X, Y, F be as in Proposition 5.1, and let Z be the set of points $y \in Y$ such that depth $F_y = n$. Then

$$\mathrm{Ass}\ \underline{H}^n_Y(F) \subseteq Z \quad ,$$

and moreover, if z is any point of Z, then some specialization of z is in $\mathrm{Ass}\ \underline{H}^n_Y(F)$. In particular, if Z has no embedded points, then

$$\mathrm{Ass}\ \underline{H}^n_Y(F) = Z \quad .$$

Proof. Let z be any point of Y, and let $Y' \subseteq Y$ be the closure of z. Then by Proposition 5.1,

$$\underline{\mathrm{Ext}}^n_{\mathcal{O}_X}(\mathcal{O}_{Y'}, F) \cong \underline{\mathrm{Hom}}_{\mathcal{O}_X}(\mathcal{O}_{Y'}, \underline{H}^n_Y(F)) \quad .$$

Therefore, using the expression for the set of associated primes of a Hom given in sublemma 3.2, we have

$$\mathrm{Ass}\ \underline{\mathrm{Ext}}^n_{\mathcal{O}_X}(\mathcal{O}_{Y'}, F) = Y' \cap \mathrm{Ass}\ \underline{H}^n_Y(F) \quad .$$

Now if $z \in \mathrm{Ass}\ \underline{H}^n_Y(F)$, then $\underline{\mathrm{Ext}}^n_{\mathcal{O}_X}(\mathcal{O}_{Y'}, F) \neq 0$, so depth $F_z = n$, and $z \in Z$. (Use Corollary 3.6 and Proposition 3.7, and observe that depth F_z can only increase when one specializes z.)

Suppose on the other hand that $z \in Z$. Then $\underline{\mathrm{Ext}}^n_{\mathcal{O}_X}(\mathcal{O}_{Y'}, F) \neq 0$, so its set of associated primes must be non-empty. If z' is one of its associated primes, then $z' \in \mathrm{Ass}\ \underline{H}^n_Y(F)$, and also $z' \in Y'$, i.e., z' is a specialization of z.

<u>Proof of Proposition 5.2</u>. Let $Y = \mathrm{Supp}\ G$. Then by Proposition 3.7, depth $F \geq n$, and, as in the proof of the lemma, we find that

$$\mathrm{Ass}\ \underline{\mathrm{Ext}}^n_{\mathcal{O}_X}(G, F) = Y \cap \mathrm{Ass}\ \underline{H}^n_Y(F) = \mathrm{Ass}\ \underline{H}^n_Y(F).$$

But since we assumed that Z has no embedded points, the lemma applies to show $\mathrm{Ass}\ \underline{H}^n_Y(F) = Z$, which completes the proof.

<u>Corollary 5.4</u>. Under the hypotheses of Proposition 5.2, $\underline{\mathrm{Ext}}^n_{\mathcal{O}_X}(G, F)$ has no embedded associated primes.

Next we study the behavior of the local cohomology groups, Ext's, and dualizing functors under base change. These results will be useful in Section 6 on duality.

Proposition 5.5. Let $f : X' \to X$ be a continuous map of topological spaces; let Y be a closed subspace of X, and let $Y' = f^{-1}(Y)$; and let F' be an abelian sheaf on X'. Then there is a spectral sequence

$$E_2^{pq} = H_Y^p(X, R^q f_*(F')) \Rightarrow H_{Y'}^n(X', F') \quad .$$

Proof. We interpret the functor $\Gamma_{Y'}$ as the composite functor $\Gamma_Y \circ f_*$, where f_* is the direct image functor. Now f_* takes injectives into injectives, since it is adjoint to the exact functor f^* [5, Ch. 0, 3.7.2]. Thus we can apply the theorem A, stated in Section 1, which gives the existence of the spectral sequence of derived functors of a composite functor.

Corollary 5.6. If in the proposition $R^q f_*(F') = 0$ for $q > 0$, then the spectral sequence degenerates, and

$$H_Y^i(X, f_* F') \cong H_{Y'}^i(X', F')$$

for all i.

Example. Let X' be a closed subprescheme of X. Then $Y' = Y \cap X'$, and for any abelian sheaf F on X with support in X', $H_Y^i(X, F) \cong H_{Y'}^i(X', F|X')$.

<u>Corollary 5.7</u>. Let $B \to A$ be a morphism of rings, let J be an ideal in B, and let M be an A-module. Then for all i we have isomorphisms

$$H^i_J(M^B) \cong H^i_{JA}(M)^B \quad ,$$

where a superscript B applied to an A-module denotes that module considered as a B-module "by restriction of scalars."

<u>Proof</u>. We apply the previous corollary to the case where $X' = \operatorname{Spec} A$ and $X = \operatorname{Spec} B$. The direct image functor f_* is exact in this case since $f : X' \to X$ is an affine morphism [5, Ch. III., Corollary 1.3.2].

<u>Remark</u>. The result of Corollary 5.7 becomes more remarkable in the presence of Noetherian assumptions when we can use Theorem 2.8 to express the groups involved as direct limits of Ext's. For then it states that there is an isomorphism

$$\varinjlim_n \operatorname{Ext}^i_B(B/J^n, M^B) \xrightarrow{\sim} \varinjlim_n \operatorname{Ext}^i_A(A/(JA)^n, M) \quad ,$$

which is surprising because the Ext groups are in general not isomorphic before taking the limit. For example, if $A = M = B/J$, the natural map

$$\operatorname{Ext}^i_B(B/J, B/J) \to \operatorname{Ext}^i_{B/J}(B/J, B/J)$$

is in general not an isomorphism for $i > 0$, since the right-hand term is 0, and the left-hand one not. However, the following proposition shows that the first non-vanishing Ext's will be isomorphic.

Proposition 5.8. Let $B \to A$ be a morphism of Noetherian rings, let N be a B-module, M an A-module, both of finite type, and let n be an integer. Then the following conditions are equivalent:

(i) $\mathrm{Ext}^i_B(N, M^B) = 0$ for all $i < n$,

(ii) $\mathrm{Ext}^i_A(N_A, M) = 0$ for all $i < n$,

where $N_A = N \otimes_B A$.

Furthermore, if the conditions are satisfied, then there is an isomorphism

$$\mathrm{Ext}^n_B(N, M^B) \xrightarrow{\sim} \mathrm{Ext}^n_A(N_A, M) \quad .$$

Proof. Let J be an ideal of B such that $\mathrm{Supp}\, N = V(J)$. Then it follows that $\mathrm{Supp}\, N_A = V(JA)$. Now by Proposition 3.7 and Theorem 3.8, the condition (i) is equivalent to saying $H^i_J(M^B) = 0$ for all $i < n$, and the condition (ii) is equivalent to saying $H^i_{JA}(M) = 0$ for all $i < n$. But these latter two conditions are equivalent by Corollary 5.7.

From Corollary 5.7 and Theorem 3.8 one also deduces that $\mathrm{depth}_J M^B = \mathrm{depth}_{JA} M$. If the conditions (i) and (ii) are satisfied, then this common depth is $\geq n$. Therefore by Proposition 5.1 there are isomorphisms

$$\mathrm{Ext}^n_B(N, M^B) \xrightarrow{\sim} \mathrm{Hom}_B(N, H^n_J(M^B))$$

and

$$\mathrm{Ext}^n_A(N_A, M) \xrightarrow{\sim} \mathrm{Hom}_A(N_A, H^n_{JA}(M)) \quad .$$

But by Corollary 5.7, again, the modules $H^n_J(M^B)$ and $H^n_{JA}(M)$ are isomorphic, hence our Ext^n's are isomorphic. (In general, if $B \to A$ is a morphism of rings; if N is a B-module of finite presentation; and if M is an A-module, then there is an isomorphism

$$\mathrm{Hom}_B(N, M^B) \xrightarrow{\sim} \mathrm{Hom}_A(N_A M)^B \quad . \quad)$$

<u>Proposition 5.9</u>. Let A be a Noetherian ring, let $J \supseteq J'$ be two ideals of A, and let M be an A-module of finite type. Denote by a hat $\hat{}$ the operation of completion with respect to the J'-adic topology. Then for all i there is an isomorphism

$$H^i_J(M) \xrightarrow{\sim} H^i_{\hat{J}}(\hat{M})^A \quad .$$

Proof. Using Theorem 2.8 to represent the local cohomology groups as direct limits of Ext's, we need only show that the natural maps

$$\mathrm{Ext}^i_A(A/J^n, M) \to \mathrm{Ext}^i_{\hat{A}}(\hat{A}/\hat{J}^n, \hat{M})$$

are isomorphisms. But since $\hat{A}$ is flat over A, and since M is of finite type, we have

$$\mathrm{Ext}^i_{\hat{A}}(\hat{A}/\hat{J}^n, \hat{M}) = \mathrm{Ext}^i_A(A/J^n, M) \otimes_A \hat{A} \quad .$$

On the other hand, the Ext group on the right is annihilated by J^n, so to tensor it with $\hat{A}$ is the same as to tensor it with $\hat{A}/J^n\hat{A}$. But $\hat{A}/J^n\hat{A} = A/J^n$, since $J \supseteq J'$, so tensoring with $\hat{A}$ is an isomorphism. This proves the proposition.

Now we study the problem of when the direct image of a coherent sheaf, under a morphism of preschemes, is coherent.

Proposition 5.10. Let X be a prescheme, let U be an open subset, let $Y = X - U$, and let F_0 be a coherent sheaf on U. Let F be any coherent sheaf on X whose restriction to U is F_0, and let n be an integer. Then the following conditions are equivalent:

(i) $R^i j_*(F_0)$ is coherent for all $i < n$, where $j : U \to X$ is the inclusion map.

(ii) $\underline{H}^i_Y(F)$ is coherent for all $i \leq n$.

Proof. By Corollary 1.9 there is an exact sequence

$$0 \to \underline{H}^0_Y(F) \to F \to j_*(F_0) \to \underline{H}^1_Y(F) \to 0 \quad ,$$

and there are isomorphisms

$$R^q j_*(F_0) \xrightarrow{\sim} \underline{H}^{q+1}_Y(F)$$

for $q > 0$. Now the statement of the proposition follows since the kernel and cokernel of a map of coherent sheaves are coherent, and an extension of a coherent sheaf by a coherent sheaf is coherent. [5, Ch. 0, 5.3]

Moreover, there is also the following result, which we will not prove.

Theorem. [16, Corollary VIII-II-3] If X is locally embeddable in a non-singular prescheme, then conditions (i) and (ii) of proposition are also equivalent to the following condition:

(iii) for all $z \in U$,

$$\text{depth } (F_0)_z + \text{codim } (Y \cap Z, Z) > n \quad ,$$

where Z is the closure of z in X.

Examples. 1) A necessary condition that $j_*(F_0)$ be coherent is that for all $z \in \mathrm{Ass}\, F_0$, $Y \cap Z$ be of codimension ≥ 2 in Z, since for $z \in \mathrm{Ass}\, F_0$, depth $(F_0)_z = 0$.

2) Suppose the conditions of proposition 5.12 satisfied, let $f : X \to Y$ be a proper morphism, where Y is locally Noetherian, and let $f_0 : U \to Y$ be the restriction of f to U. Then the sheaves $R^q f_{0*}(F_0)$ are coherent on Y for all $i < n$.

Indeed, we can interpret f_{0*} as the composite functor $f_* \circ j_*$, and j_* carries injectives into injectives, so there is a spectral sequence

$$E_2^{pq} = R^p f_*(R^q j_*(F_0)) \Rightarrow R^i f_{0*}(F_0) \quad .$$

Now the sheaves $R^q j_*(F_0)$ are coherent for $q < n$, by the proposition. Thus the E_2^{pq} terms are coherent sheaves for $q < n$ and for all p, since f is a proper map [5, Ch. III, Theorem 3.2.1]. It follows that the abutment terms $R^i f_{0*}(F_0)$ are coherent for $i < n$!

Problem.* Let X be a prescheme, and let Y be a closed subset. If n is an integer, find conditions under which $\underline{H}^i_Y(F) = 0$ for all $i > n$, and for all quasi-coherent sheaves F.

It is sufficient that Y be locally describable by n equations, but not necessary. For example, if X is the cone

*This problem is studied in [19].

over a non-singular plane cubic curve, and if Y is a generator of the cone, then the condition is satisfied for $n = 1$, but Y cannot necessarily be described by one equation at the vertex of the cone.

§ 6. Local Duality

In this section we prove a central theorem of the local cohomology theory, the duality theorem. It is simplest to state over a regular local ring (Theorem 6.3); however, we also give forms of the duality theorem valid over more general local rings. These theorems should be thought of as local analogues of Serre's projective duality theorem [10, n°72, Theorem 1], which says that if F is a coherent sheaf on projective r-space $X = \mathbb{P}^r$, then for $i \geq 0$, there is a perfect pairing

$$H^i(X;F) \times \mathrm{Ext}^{r-i}_{\mathcal{O}_X}(F, \Omega) \to k$$

of finite-dimensional vector spaces over k, where Ω is the sheaf of r-differential forms on $\mathbb{P}^r$, isomorphic to $\mathcal{O}_X(-r-1)$. Even the proof, which uses the axiomatic characterization of derived functors, is similar.

We will apply the duality theorem to determine the structure of the local cohomology groups $H^i_{\mathfrak{m}}(M)$ over a local ring, and we end by giving an application of duality to a theorem on algebraic varieties.

We begin by recalling briefly Yoneda's interpretation of the Ext groups (for details and proofs, see [7, exposé 3]).

Let $\mathcal{C}$ be an abelian category with enough injectives. A complex in $\mathcal{C}$ is a collection $(K^n)_{n \in \mathbb{Z}}$ of objects of $\mathcal{C}$, together with coboundary maps $d^n : K^n \to K^{n+1}$ such that for all n, $d^{n+1} \circ d^n = 0$. We denote the cohomology of a complex K by $H^q(K)$.

If K, L are two complexes in $\mathcal{C}$, a morphism of K into L of degree s is a collection $(f^n)_{n \in \mathbb{Z}}$ of morphisms $f^n : K^n \to L^{n+s}$ which commute with the coboundary maps in the two complexes. We denote by $\mathrm{Hom}^s(K, L)$ the group of morphisms of K into L of degree s. Two such morphisms f, g are said to be homotopic if there exists a collection $(p^n)_{n \in \mathbb{Z}}$ of morphisms $p^n : K^n \to L^{n+s-1}$ such that for all n,

$$f^n - g^n = d_L^{n+s-1} \circ p^n + (-1)^s p^{n+1} \circ d_K^n \quad .$$

We denote by $\underset{\sim}{H}^s(K, L)$ the group of homotopy classes of elements $f \in \mathrm{Hom}^s(K, L)$.

If A is an object of $\mathcal{C}$, a resolution of A is a complex K, such that $K^n = 0$ for $n < 0$, together with a map $\epsilon : A \to K^0$, such that the sequence

$$0 \to A \xrightarrow{\epsilon} K^0 \to K^1$$

is exact. A resolution (K, ϵ) of A is said to be exact if

$H^q(K) = 0$ for $q > 0$. The resolution is said to be injective if each K^i is injective.

Now let A be an object of $\mathcal{C}$, let (K, ϵ) be a resolution of A, and let L be any complex. We denote by $\text{Hom}(A, L)$ the complex $(\text{Hom}(A, L^n))_{n \in \mathbb{Z}}$. Given an $f \in \text{Hom}^s(K, L)$, we define an element of $\text{Hom}(A, L)^s$ by composing ϵ with f^0. One shows that this element is a cocycle in $\text{Hom}(A, L)$, and hence determines an element of $H^s(\text{Hom}(A, L))$, which depends only on the homotopy class of f. Thus we have defined a natural map for each s,

$$\phi^s : \underset{\sim}{H}^s(K, L) \to H^s(\text{Hom}(A, L)) .$$

Theorem E. If (K, ϵ) is an exact resolution of A, and if the complex L consists entirely of injectives, then the natural maps ϕ^s defined above are isomorphisms.

Corollary. If in particular B is another object of $\mathcal{C}$, and L is an exact injective resolution of B, then there are canonical isomorphisms

$$\underset{\sim}{H}^s(K, L) \xrightarrow{\sim} \text{Ext}^s(A, B) .$$

Now we apply this theory to the situation we will meet below.

<u>Proposition 6.1</u>. Let $\mathcal{C}$, $\mathcal{C}'$ be abelian categories, $\mathcal{C}$ having enough injectives, and let $T : \mathcal{C} \to \mathcal{C}'$ be an additive covariant left-exact functor. If A and B are objects of $\mathcal{C}$, then there are pairings

$$R^iT(A) \times \mathrm{Ext}^s(A,B) \to R^{i+s}T(B)$$

for all i, s.

<u>Proof</u>. Pick exact injective resolutions (K, ϵ) and (L, η) of A and B, respectively. Then we can calculate $R^iT(A)$ and $R^iT(B)$ as the cohomology of the complexes T(K) and T(L). If f is a morphism of K into L of degree s, then T(f) is a morphism of T(K) into T(L) of degree s, which on passing to cohomology gives a morphism

$$f_* : R^iT(A) \to R^{i+s}T(B)$$

for any s. One sees easily that f_* depends only on the homotopy class of f. Thus using the isomorphism of the corollary above, we have defined a pairing

$$R^iT(A) \times \mathrm{Ext}^s(A, B) \to R^{i+s}\,T(B)$$

for all i, s.

Corollary 6.2. Let A be a ring, let $\mathfrak{m}$ be an ideal, and let M, N be A-modules. Then there are pairings

$$H^i_{\mathfrak{m}}(M) \times \mathrm{Ext}^j_A(M,N) \to H^{i+j}_{\mathfrak{m}}(N)$$

for all i, j.

Proof. Let T be the functor $M \rightsquigarrow H^0_{\mathfrak{m}}(M)$ from the category of A-modules into itself, and apply the previous proposition. The $H^i_{\mathfrak{m}}(M)$ for $i > 0$ are right derived functors of T by Corollary 2.7.

Theorem 6.3. (Duality) Let A be a Gorenstein ring of dimension n (See proposition 4.13), let $\mathfrak{m}$ be the maximal ideal, let $I = H^n_{\mathfrak{m}}(A)$ be a dualizing module, and let D be the functor $\mathrm{Hom}(\cdot, I)$. Let M be a module of finite type. Then the pairing

$$(*) \qquad H^i_{\mathfrak{m}}(M) \times \mathrm{Ext}^{n-i}_A(M,A) \to I$$

gives rise to isomorphisms

$$\phi_i : H^i_{\mathfrak{m}}(M) \xrightarrow{\sim} D(\mathrm{Ext}^{n-i}_A(M,A))$$

and

$$\psi_i : \mathrm{Ext}^{n-i}_A(M,A)^{\wedge} \xrightarrow{\sim} D(H^i_{\mathfrak{m}}(M)) \quad .$$

In particular, if A is complete, then (*) is a perfect pairing. (A pairing $L \times M \to N$ is said to be <u>perfect</u> if the maps $L \to \mathrm{Hom}(M, N)$ and $M \to \mathrm{Hom}(L, N)$ it induces are both isomorphisms.)

<u>Proof</u>. We first consider the case $i = n$, and show that the map

$$\phi_n : H^n_{\mathfrak{m}}(M) \to D(\mathrm{Hom}(M, A))$$

is an isomorphism. Indeed, ϕ_n is an isomorphism for $M = A$, since $H^n_{\mathfrak{m}}(A) = I = D(A)$. Considered as functors in M, both $H^n_{\mathfrak{m}}(M)$ and $D(\mathrm{Hom}(M, A))$ are right-exact and covariant: Hence by a now familiar argument, ϕ_n is an isomorphism for all M of finite type.

To show that the remaining ϕ_i are isomorphisms, we use the axiomatic characterization of derived functors. Since D is exact, the functors $D(\mathrm{Ext}^{n-i}_A(\cdot\, , A))$ for $i < n$ are left-derived functors of $D(\mathrm{Hom}(\cdot\, , A))$. The functors $H^i_{\mathfrak{m}}(\cdot)$ form a connected sequence of functors, and ϕ_n is an isomorphism. Hence, to show that the other ϕ_i are isomorphisms, we need only show that the functors $H^i_{\mathfrak{m}}(\cdot)$ for $i < n$ are left-derived functors of $H^n_{\mathfrak{m}}(M)$, i.e., that $H^i_{\mathfrak{m}}(P) = 0$ for P projective and $i < n$. In fact, since A is Noetherian and since we are interested only in M of finite type, we may assume P is of finite type; since $H^i_{\mathfrak{m}}(\cdot)$ is additive, we need only show that $H^i_{\mathfrak{m}}(A) = 0$ for $i < n$. This follows from the fact that A is Cohen-Macaulay. Thus ϕ_i is an isomorphism for all i.

Applying the dualizing functor D to the isomorphisms ϕ_i, we obtain the isomorphisms

$$D(\phi_i) : DD(\mathrm{Ext}_A^{n-i}(M,A)) \xrightarrow{\sim} D(H^i_{\mathfrak{m}}(M)) \quad .$$

But $DD(M) = \hat{M}$ for any module M of finite type. (Indeed, the functors $M \rightsquigarrow DD(M)$ and $M \rightsquigarrow \hat{M}$ are both right exact covariant; they are isomorphic for $M = A$, hence they are isomorphic for all M of finite type.) Hence we have also the isomorphisms ψ_i.

Q.E.D.

We now consider an arbitrary Noetherian local ring A (subject however to the weak restriction that it be a quotient of a regular local ring) and prove two propositions which describe the structure of the modules $H^i_{\mathfrak{m}}(M)$, where M is a module of finite type, and $\mathfrak{m}$ is the maximal ideal. Most of these properties are easier to express in terms of the duals, so for each i we define $T^i(M) = D(H^i_{\mathfrak{m}}(M))$, where D is some fixed dualizing functor. Then the T^i are contravariant additive functors in M; T^0 is right exact, and the T^i for $i > 0$ are the left derived functors of T^0.*

Proposition 6.4. Let A be a Noetherian local ring (quotient of a regular local ring) of arbitrary dimension, and let M be an A-module of finite type and of dimension $n \geq 0$. Then

1) $H^i_{\mathfrak{m}}(M) = 0$ and $T^i(M) = 0$ for $i > n$.

2) The modules $T^i(M)$ are of finite type over $\hat{A}$.

3) For each i, $\dim_{\hat{A}} T^i(M) \leq i$.

4) $\dim_{\hat{A}} T^n(M) = n$. (In particular, $H^n_{\mathfrak{m}}(M) \neq 0$.)

Proof. 1) M is also a module over the local ring A/Ann M, which is of dimension n. Thus by Corollary 5.7 we may assume that A is of dimension n. Then we see that $H^i_{\mathfrak{m}}(M) = 0$ for $i > n$ either by using Proposition 1.12, or by expressing $H^i_{\mathfrak{m}}(M)$ in terms of a limit of Koszul complexes of n elements, by Theorem 2.3.

For the remaining statements, fix a regular local ring B of some dimension r of which A is a quotient. Then $B \to A$ is a surjective map of Noetherian local rings, and using Corollary 5.7

*In fact, the functors T^i, which really have nothing to do with local cohomology, can be interpreted as $\mathrm{Ext}^i(\cdot, R^\cdot)$ for a suitable dualizing complex $R^\cdot$ on A. See [17, ch V § 2].

and Proposition 4.12 we see that $T^i(M)$ is the same whether calculated over A or over B. Thus we may apply the duality theorem 6.3, and find

$$T^i(M) \cong \mathrm{Ext}_B^{r-i}(M,B)\hat{} \cong \mathrm{Ext}_{\hat{B}}^{r-i}(\hat{M},\hat{B}) \quad .$$

To prove 2) observe that since M is of finite type over A, it is also over B, and hence $\hat{M}$ is of finite type over $\hat{B}$. But an Ext of modules of finite type is again of finite type, so $T^i(M)$ is of finite type over $\hat{B}$ for all i. But it is also a module over $\hat{A}$, hence it is of finite type over $\hat{A}$.

To prove 3), we note that $\hat{B}$ is also a regular local ring, and that the dimension of $T^i(M)$ is the same, whether calculated over $\hat{A}$ or $\hat{B}$. Thus we have reduced the problem to the following:

<u>Lemma 6.5</u>. Let A be a regular local ring of dimension r; let M and N be modules of finite type over A. Then for all i,

$$\dim \mathrm{Ext}_A^{r-i}(M,N) \leq i \quad .$$

<u>Proof</u>. To say that an A-module L is of dimension $\leq i$ is to say that for any prime $\mathfrak{p} \subseteq A$ of coheight $> i$, the module $L_{\mathfrak{p}}$ is zero. If $\mathfrak{p}$ is such a prime, then $A_{\mathfrak{p}}$ is a regular

local ring of dimension $< r - i$. By Serre's theorem its global homological dimension is $< r - i$, so

$$\mathrm{Ext}_A^{r-i}(M, N)_{\mathfrak{p}} = \mathrm{Ext}_{A_{\mathfrak{p}}}^{r-i}(M_{\mathfrak{p}}, N_{\mathfrak{p}}) = 0 .$$

Thus $\dim \mathrm{Ext}_A^{r-i}(M, N) \leq i$.

To prove assertion 4) of the proposition, we have

$$T^n(M) \cong \mathrm{Ext}_{\hat{B}}^{r-n}(\hat{M}, \hat{B}) .$$

By assertion 1) which we have already proved, this is a first non-vanishing Ext group. Applying Proposition 5.2, we find that Z is just the set of points $y \in \mathrm{Supp}\ \hat{M}$ whose dimension is n : since $\hat{B}$ is regular, depth $\hat{B}_y = \dim \hat{B}_y = \dim y$ for any $y \in \mathrm{Spec}\ \hat{B}$. In particular, Z has no embedded points, so

$$\mathrm{Ass}_{\hat{A}}\ T^n(M) = Z ,$$

which shows that $\dim_{\hat{A}} T^n(M) = n$. But furthermore, since $\dim_{\hat{A}} \hat{M} = n$, every point of Z is associated to $\hat{M}$, so Z is the set of points of $\mathrm{Ass}_{\hat{A}} \hat{M}$ of dimension n. This proves assertion 5) of the next proposition.

Notation. If $\mathfrak{a}$ is an ideal of dimension n in a Noetherian ring A, we will denote by $\mathfrak{a}_n$ the intersection of those primary ideals in a Noetherian decomposition of $\mathfrak{n}$ which are of dimension n. (These primary ideals all belong to minimal primes of $\mathfrak{a}$, hence are uniquely determined.) If Z is a set of prime ideals of A, we will denote by Z_n those primes in Z which are of dimension n.

Proposition 6.6. Let A be a complete local Noetherian ring (quotient of a regular local ring) of arbitrary dimension, and let M be an A-module of finite type of dimension $n \geq 0$. Then

5) $\mathrm{Ass}\ T^n(M) = (\mathrm{Ass}\ M)_n$

6) If $\mathfrak{p} \in \mathrm{Ass}\ T^n(M)$, then

$$\ell_{\mathfrak{p}}(T_n(M)) = \ell_{\mathfrak{p}}(M) ,$$

where for any A-module N, $\ell_{\mathfrak{p}}(N)$ denotes the length of $N_{\mathfrak{p}}$ over the local ring $A_{\mathfrak{p}}$.

7) $\mathrm{Ann}\ T^n(M) = (\mathrm{Ann}\ M)_n$.

8) There is a natural map

$$\alpha : M \to T^n T^n(M)$$

whose kernel is the set of elements of M whose support has dimension $< n$.

<u>Proofs</u>. 5) was already proved above. To prove 6), let $\mathfrak{p} \in \operatorname{Ass} T^n(M)$. Then, letting B be a complete regular local ring of dimension r of which A is a quotient,

$$T^n(M) \cong \operatorname{Ext}_B^{r-n}(M, B)$$

and so

$$T^n(M)_{\mathfrak{p}} \cong \operatorname{Ext}_{B_{\mathfrak{p}}}^{r-n}(M_{\mathfrak{p}}, B_{\mathfrak{p}}) \quad .$$

But by 5), $\mathfrak{p}$ is of dimension n, so $B_{\mathfrak{p}}$ is a regular local ring of dimension $r - n$. Hence by Proposition 4.13, the functor

$$M_{\mathfrak{p}} \rightsquigarrow \operatorname{Ext}_{B_{\mathfrak{p}}}^{r-n}(M_{\mathfrak{p}}, B_{\mathfrak{p}})$$

is dualizing, and so preserves lengths and annihilators. This proves 6) and also 7), since the primes in $\operatorname{Ass} T^n(M)$ are the only ones which could be associated to $\operatorname{Ann} T^n(M)$.

8) $T^n(M)$ is a left exact contravariant functor in M, so by Proposition 4.2 there is a canonical isomorphism

$$T^n(M) \xrightarrow{\sim} \operatorname{Hom}_A(M, \Omega) \quad ,$$

where $\Omega = T^n(A)$. By means of this isomorphism we can define the natural map α. If x is an element of M whose support has dimension $< n$, then any morphism $f : M \to \Omega$ annihilates x, since by 5) above, $\operatorname{Ass} \Omega$ is of pure dimension n. Hence $x \in \ker \alpha$.

Conversely, let x be an element of M whose support is of dimension n. To show that $\alpha(x) \neq 0$, it will be sufficient to show that there is a morphism $f : M \to \Omega$ such that $f(x) \neq 0$. Or, equivalently, letting N be the submodule of M generated by x, it will be sufficient to show that the natural map

$$T^n(j) : T^n(M) \to T^n(N) \quad ,$$

where $j : N \to M$ is the canonical injection, is not zero. For then there will be an $f \in T^n(M) = \mathrm{Hom}_A(M, \Omega)$ such that f restricted to N is not zero, i.e., $f(x) \neq 0$.

From the exact sequence,

$$0 \to N \xrightarrow{j} M \to M/N \to 0 \quad ,$$

we deduce an exact sequence

$$T^n(M) \xrightarrow{T^n(j)} T^n(N) \to T^{n-1}(M/N) \quad .$$

But by 3) above, $\dim T^{n-1}(M/N) \leq n-1$, and by 4), $\dim T^n(N) = n$, since N is of dimension n. Therefore, $T^n(j)$ must be non-zero, since if it were zero, $T^{n-1}(M/N)$ would contain an isomorphic copy of $T^n(N)$ which is impossible because of their dimensions. Hence $\ker \alpha$ is just those elements of M whose support has dimension $< n$.

<u>Example</u>. If A is a complete local domain of dimension n, then $\Omega = T^n(A)$ is torsion-free of rank 1. Indeed, by 5), its only associated prime is the zero ideal, hence it is torsion-free.

Moreover, taking $\mathfrak{J}$ to be the zero ideal in A, $\ell_{\mathfrak{J}}(N)$ is just the rank of N, for any A-module N. Hence rank Ω = rank A = 1.

<u>Definition</u>. Let A be a complete local ring of dimension n, let D be a dualizing functor, and let $\Omega = T^n(A) = D(H^n_{\mathfrak{m}}(A))$. Then Ω is called a <u>module of dualizing differentials</u> for A. It is determined up to isomorphism.

<u>Exercise</u>. If M is a Cohen-Macaulay module, show that the map α of 8) above is an isomorphism. In particular, $A \xrightarrow{\alpha} T^n T^n(A) = \mathrm{Hom}_A(\Omega, \Omega)$ is an isomorphism.

<u>Problem</u>.* Let A be a Noetherian ring, J an ideal in A, and M a module of finite type. Consider the local cohomology modules

$$H^i_J(M) = \varinjlim_n \mathrm{Ext}^i_A(A/J^n, M) \quad .$$

Find finiteness statements of an Artin-Rees type, to generalize assertion 2) of Proposition 6.4. But they should be stated without using duality. One should also have commutativity with projective limits (cf. the invariance theorem of Zariski, and the Lefschetz relations of completions along a subvariety).

---

*This problem is studied in [18]. See also [16, exposé XIII] for a more precise statement of the problem.

Now we come to duality theorems for non-regular local rings. For a complete Cohen-Macaulay local ring, it will be sufficient to replace $\mathrm{Ext}^{n-i}(M,A)$ in the statement of Theorem 6.3 by $\mathrm{Ext}^{n-i}(M,\Omega)$, where Ω is a module of dualizing differentials. If our local ring is not Cohen-Macaulay, however, there will be a finite number of modules $\Omega = \Omega^0, \Omega^1, \ldots, \Omega^n$ determined up to isomorphism by the local ring, and we replace $\mathrm{Ext}^{n-i}(M,A)$ in Theorem 6.3 by the abutment of a spectral sequence whose E_2^{pq} terms are $\mathrm{Ext}^p(M, \Omega^q)$.

Let A be a complete Noetherian local ring of dimension n (quotient of a regular local ring). Fix a dualizing module I and the associated dualizing functor D, and let $\Omega = D(H^n_{\mathfrak{m}}(A))$ be a module of dualizing differentials for A. Then Ω is of finite type over A by 2) of Proposition 6.4, and is torsion-free of rank 1 if A is a domain. Since the functor $H^n_{\mathfrak{m}}(\cdot)$ is right exact on the category of A-modules, there is a canonical isomorphism

$$H^n_{\mathfrak{m}}(A) \otimes_A M \xrightarrow{\sim} H^n_{\mathfrak{m}}(M)$$

for all M of finite type (the usual argument ...). Therefore,

$$H^n_{\mathfrak{m}}(\Omega) \cong H^n_{\mathfrak{m}}(A) \otimes_A \mathrm{Hom}_A(H^n_{\mathfrak{m}}(A), I) \quad ,$$

and we see that there is a natural map

$$H^n_{\mathfrak{m}}(\Omega) \to I \quad .$$

Composing this map with the pairings of Corollary 6.2, we can define pairings for all i

$$(*) \qquad H^i_{\mathfrak{m}}(M) \times \mathrm{Ext}^{n-i}_A(M, \Omega) \to I \quad .$$

Theorem 6.7. The following conditions are equivalent (where $k \geq 0$ is an integer):

(i) the pairings $(*)$ are perfect for all $n - k \leq i \leq n$.

(ii) $H^i_{\mathfrak{m}}(A) = 0$ for all $n - k \leq i < n$.

In particular, the pairings $(*)$ are perfect for all i if and only if A is Cohen-Macaulay.

First Proof. We mimic the proof of Theorem 6.3. We first consider the case $i = n$, and the morphism

$$\phi_n : H^n_{\mathfrak{m}}(M) \to D(\mathrm{Hom}\,(M, \Omega)) \quad .$$

This is an isomorphism for $M = A$, since $D(\Omega) = DD(H^n_{\mathfrak{m}}(A)) = H^n_{\mathfrak{m}}(A)$. Moreover, both sides are right exact covariant functors in M, so ϕ_n is an isomorphism for all M of finite type.

To show that

$$\phi_i : H^i_{\mathfrak{m}}(M) \to D(\mathrm{Ext}^{n-i}_A(M, \Omega))$$

is an isomorphism for $n - k \leq i < n$, we need only show that the functors $H^i_{\mathfrak{m}}(\cdot)$ are left derived functors of $H^n_{\mathfrak{m}}(\cdot)$, i.e., that $H^i_{\mathfrak{m}}(A) = 0$ for $n - k \leq i < n$. This is precisely our condition (ii). That the pairing is perfect for $n - k \leq i \leq n$ follows as in the proof of Theorem 6.3.

Conversely, assume condition (i). Then $H^i_{\mathfrak{m}}(A)$, for $n - k \leq i \leq n$, is dual to $\mathrm{Ext}^{n-i}_A(A, \Omega) = 0$ for $i < n$. Therefore, $H^i_{\mathfrak{m}}(A) = 0$ for $n - k \leq i < n$.

The last statement follows from Theorem 3.8, since A is Cohen-Macaulay if and only if its depth is n.

<u>Second Proof</u>. Instead of proving the theorem directly, we deduce it from Theorem 6.3. Let B be a complete regular local ring of dimension r of which A is a quotient. Let J be a dualizing module for B (we may assume, by Proposition 4.12, that $I = \mathrm{Hom}_B(A, J) \subseteq J$). Then by Theorem 6.3 and Corollary 5.7 there is a perfect pairing

(1) $$H^i_{\mathfrak{m}}(M) \times \mathrm{Ext}^{r-i}_B(M, B) \to J$$

for all i, and for all M of finite type over A (or B). If M is an A-module, then both partners to this pairing are A-modules, so the image of the pairing will be in I. (In fact, I is the largest sub-B-module of J which is also an A-module.)

Thus to obtain a duality statement intrinsic over A, we need only express the modules $\mathrm{Ext}_B^{r-i}(M, B)$ in terms of things defined intrinsically over A.

There is a spectral sequence associated with a change of rings [12, Ch. XVI, § 5]:

$$E_2^{pq} = \mathrm{Ext}_A^p(M, \mathrm{Ext}_B^q(A, B)) \Rightarrow \mathrm{Ext}_B^i(M, B) \quad .$$

This reduces the problem to calculating the groups $\mathrm{Ext}_B^q(A, B)$, which (except for their numbering) do not depend on the choice of B, since they are dual to $H_{\mathfrak{m}}^{r-q}(A)$! In our situation we have

$$\mathrm{Ext}_B^q(A, B) = \begin{cases} 0 & \text{for } q < r - n \\ \Omega & \text{for } q = r - n \\ \cdots & \\ 0 & \text{for } q > r \end{cases} \quad .$$

Assuming condition (ii) of the theorem, $\mathrm{Ext}_B^q(A, B) = 0$ for $r - n < q \leq r - n + k$. Therefore, the spectral sequence degenerates partially, and we find

$$\mathrm{Ext}_B^{r-i}(M, B) \cong \mathrm{Ext}_A^{n-i}(M, \Omega)$$

for $i \geq n - k$. Substituting in (1) gives condition (i) of the theorem.

The converse (i) => (ii) is proved as in the first proof.

Now we come to the most general duality theorem, whose proof will generalize the second proof of the last theorem.

Theorem 6.8. Let A be a complete local ring of dimension n (quotient of a regular local ring). Let $\Omega^i = D(H^{n-i}_{\mathfrak{m}}(A))$ for $i = 0, 1, \ldots, n$. Then there is a spectral sequence

$$E_2^{pq} = \operatorname{Ext}^p_A(M, \Omega^q) \Rightarrow E^i = D(H^{n-i}_{\mathfrak{m}}(M)) \quad .$$

Proof. As in the second proof of Theorem 6.7 above, let B be a complete regular local ring of dimension r, of which A is a quotient. Then by Theorem 6.3,

$$\operatorname{Ext}^{r-i}_B(M, B) \cong D(H^i_{\mathfrak{m}}(M))$$

for all i. Furthermore, there is a spectral sequence (see above),

$$E_2^{pq} = \operatorname{Ext}^p_A(M, \operatorname{Ext}^q_B(A, B)) \Rightarrow \operatorname{Ext}^i_B(M, B) \quad .$$

Since $\operatorname{Ext}^q_B(A, B)$ is dual to $H^{r-q}(A)$, we have

$$\operatorname{Ext}^i_B(A, B) = \begin{cases} 0 & \text{for } i < r-n \\ \Omega^q & \text{for } i = r-n+q \\ 0 & \text{for } i > r \quad . \end{cases}$$

Now by substituting and juggling the indices one obtains the statement of the theorem.

Remark. One can give another interpretation of the groups $\mathrm{Ext}^i_B(M,B)$, intrinsic over A, as follows. Let $K(B)$ be an exact injective resolution of B in the category of B-modules. Let K_A be the complex $\mathrm{Hom}_B(A; K(B))$, shifted $r-n$ places to the left. Then K_A is an injective complex over A, uniquely determined to within homotopy,* and

$$H^i(K_A) = \begin{cases} 0 & \text{for } i < 0 \\ \Omega^i & \text{for } i \geq 0 \end{cases} .$$

(In particular, K_A is an injective resolution of $\Omega = \Omega^0$ if and only if A is Cohen-Macaulay.) Now

$$\mathrm{Ext}^{r-i}_B(M,B) = H^{n-i}(\mathrm{Hom}_A(M,K_A)) \quad ,$$

which is, by definition, the hyperext group $\underline{\underline{\mathrm{Ext}}}^{n-i}_A(M,K_A)$.

As an application of the theory of duality, we prove the following theorem, conjectured by Lichtenbaum.

Theorem 6.9. Let X be a quasi-projective scheme over a field k, of dimension n. Then for any coherent sheaf F on X, $H^n(X,F)$ is a finite-dimensional vector space over k. Furthermore, the following conditions are equivalent:

*K_A is a dualizing complex for A, in the sense of [17, ch.V § 2].

(i) All irreducible components of X of dimension n are non-proper.

(ii) $H^n(X, F) = 0$ for all quasi-coherent sheaves F on X.

(iii) $H^n(X, \mathcal{O}_X(-m)) = 0$ for all $m >> 0$, where $\mathcal{O}_X(1)$ is a very ample sheaf induced by some projective embedding.

<u>Proof.</u>* Since X is quasi-projective, we can embed it as a locally closed subscheme of some projective space $\mathbb{P}^r$. Let X' be its closure in $\mathbb{P}^r$. Then any coherent sheaf F on X is the restriction of a coherent sheaf F' on X', and there is an exact sequence

$$H^n(X', F') \to H^n(X, F) \to H^{n+1}_{X'-X}(X', F') \quad .$$

The last term is zero, by Proposition 1.12, and the first is a finite-dimensional vector space over k [10, n°66, Theorem 1], so the middle term $H^n(X, F)$ is also. This proves the first assertion.

(i) => (ii). Again let X' be a projective completion of X. Let Y be a finite set of closed points, one from each irreducible component of X', and such that $X \subseteq X' - Y \subseteq X'$. Then $X' - Y$ also satisfies condition (i), and one sees immediately that it is sufficient to treat the case $X = X' - Y$.

---

*Another proof, not using local cohomology, has since been given by Kleiman [20], and a third proof, using purely local methods, is given in [19].

Since H^n commutes with direct limits (Proposition 2.11) it is sufficient to prove (ii) for coherent sheaves F. There is an exact sequence

$$H^n_Y(X', F) \xrightarrow{\alpha} H^n(X', F) \to H^n(X, F) \to 0 \quad ,$$

where F is any coherent sheaf on X'. To show $H^n(X, F) = 0$, is the same as showing α surjective, or equivalently, that the map

$$\mathrm{Hom}_k(H^n_Y(X', F), k) \xleftarrow{\alpha^*} \mathrm{Hom}_k(H^n(X', F), k)$$

of the dual vector spaces is injective. We interpret these vector spaces by means of local and global duality theorems. Let X' be embedded in projective space $V = \mathbb{P}^r$, and let Ω be the sheaf of r-differential forms on V. Then by Serre's global duality theorem,

$$\mathrm{Hom}_k(H^n(X', F), k) \cong \mathrm{Ext}^{r-n}_{\mathcal{O}_V}(F, \Omega) \quad .$$

For the local cohomology group, observe that by the excision formula (Proposition 1.3),

$$H^n_Y(X', F) = H^n_Y(\coprod_i \mathrm{Spec}\, \mathcal{O}_{y_i}, F) = \coprod_i H^n_{\mathfrak{m}_i}(F_{y_i}) \quad ,$$

where $Y = \{y_i\}$, and where $\mathcal{O}_{y_i}$ is the local ring of y_i in V; $\mathfrak{m}_i$ its maximal ideal, and F_{y_i} the stalk of F at y_i. Now by the local duality theorem 6.3, and the fact that $\mathrm{Hom}_k(\cdot, k)$ is a dualizing functor for modules with support in $V(\mathfrak{m}_i)$ (See Example 3 of §4)

$$\mathrm{Hom}_k(H^n_Y(X', F), k) \cong \coprod_i \mathrm{Ext}^{r-n}_{\hat{\mathcal{O}}_{y_i}}(\hat{F}_{y_i}, \hat{\Omega}_{y_i}) .$$

Moreover (and the justification of this statement* requires the examination of the residue map, and of the relation between local and global duality), the transported map

$$\coprod_i \mathrm{Ext}^{r-n}_{\hat{\mathcal{O}}_{y_i}}(\hat{F}_{y_i}, \hat{\Omega}_{y_i}) \overset{\alpha^*}{\leftarrow} \mathrm{Ext}^{r-n}_{\mathcal{O}_V}(F, \Omega)$$

is the one induced on the derived functors by the natural map (for any two sheaves F, G on V)

$$\mathrm{Hom}_{\mathcal{O}_V}(F, G) \to \mathrm{Hom}_{\mathcal{O}_{y_i}}(F_{y_i}, G_{y_i}) \to \mathrm{Hom}_{\hat{\mathcal{O}}_{y_i}}(\hat{F}_{y_i}, \hat{G}_{y_i})$$

To show that α^* is injective, observe that $\mathrm{Ext}^{r-n}_{\mathcal{O}_V}(F, \Omega)$ is the ending of a spectral sequence (see Proposition 2.10) whose E_2^{pq}-term is

$$E_2^{pq} = H^p(V, \underline{\mathrm{Ext}}^q_{\mathcal{O}_V}(F, \Omega)) .$$

---

*The compatibility of local and global duality is proved in [17, ch VIII, Prop. 3.5].

But since the support of F is contained in X' of dimension n, $\underline{\mathrm{Ext}}^q_{\mathcal{O}_V}(F,\Omega) = 0$ for $q < r-n$, so

$$\mathrm{Ext}^{r-n}_{\mathcal{O}_V}(F,\Omega) \cong H^0(V, \underline{\mathrm{Ext}}^{r-n}_{\mathcal{O}_V}(F,\Omega)) .$$

Now $\underline{\mathrm{Ext}}^{r-n}_{\mathcal{O}_V}(F,\Omega)$ is either zero, in which case there is nothing to prove, or a first non-vanishing Ext. In the latter case, any global section has support of pure dimension n by Proposition 5.2: thus if it vanishes at all the points y_i under the map α^*, it must be zero. Thus α^* is injective.

(ii) <=> (iii). One implication is obvious. The other follows from the fact that for F coherent and for all m large enough, one can find surjections

$$\mathcal{O}_X(-m)^p \to F \to 0 .$$

(iii) => (i). One sees easily that it is sufficient to show that if X is projective and irreducible of dimension n over k, then $H^n(X, \mathcal{O}_X(-m)) \neq 0$ for all $m >> 0$. Embedding X in $V = \mathbb{P}^r$, this H^n is dual to

$\mathrm{Ext}^{r-n}_{\mathcal{O}_V}(\mathcal{O}_X(-m), \Omega) = H^0(V, \underline{\mathrm{Ext}}^{r-n}_{\mathcal{O}_V}(\mathcal{O}_X, \Omega)(m))$. But this sheaf $\underline{\mathrm{Ext}}$ is non-zero by Proposition 5.2, hence for all m large enough its global sections are non-zero.

Q.E.D.

Bibliography

1. M. Auslander and D. Buchsbaum: "Codimension and Multiplicity," Annals of Mathematics, vol. 68 no. 3 (1958) pp. 625-657.

2. Eckmann and Schopf: "Ueber injektive Moduln," Archiv der Mathematik 4 (1953).

3. P. Gabriel: "Des Catégories Abéliennes," Bull. Soc. Math. France 90 (1962) 323-448.

4. R. Godement: "Topologie Algébrique et Théorie des Faisceaux," Hermann, Paris (1958).

5. A. Grothendieck: "Eléments de Géométrie Algébrique," Institut des Hautes Etudes Scientifiques, Paris (1960, ff.)

6. A. Grothendieck: "Sur Quelque Points d'Algèbre Homologique," Tohoku Mathematical Journal, vol. IX (1957) pp. 119-221.

7. A. Grothendieck: "Seminaire d'Algèbre Homologique," mimeographed notes, Paris (1957).

8. R. Hartshorne: "Complete Intersections and Connectedness," American Journal of Mathematics, 84 (1962) 497-508.

9. E. Matlis: "Injective Modules over Noetherian Rings," Pacific Journal of Mathematics, vol. 8 no. 3 (1958) pp. 511-528.

10. J.-P. Serre: "Faisceaux Algébriques Cohérents," Annals of Mathematics, vol. 61 no. 2 (1955) pp. 197-278.

11. J.-P. Serre: "Algèbre Locale--Multiplicités," Springer-Verlag, "Lecture Notes in Mathematics," no. 11 (1965).

12. H. Cartan and S. Eilenberg: "Homological Algebra," Princeton University Press (1956).

(Bibliography, continued)

13. H. Bass, "On the ubiquity of Gorenstein rings," Math. Zeitschr. 82 (1963) 8-28.

14. Séminaire Cartan, 1948-49, "Séminaire de topologie algébrique," mimeo notes, Secr. Math. Paris.

15. Séminaire Cartan, 1950-51, "Cohomologie des Groupes, Suites spectrales, Faisceaux," mimeo notes, Secr. Math. Paris.

16. A. Grothendieck, "Séminaire de Geométrie Algébrique 1962," mimeo notes, I.H.E.S. Paris. (3rd ed. to appear, North Holland Pub. Co.)

17. R. Hartshorne, "Residues and Duality," Springer Verlag, "Lecture Notes in Mathematics," no. 20 (1966).

18. R. Hartshorne, "Affine duality and cofiniteness," to appear.

19. R. Hartshorne, "Cohomological dimension of algebraic varieties," to appear.

20. S. Kleiman, "On the vanishing of $H^n(X,F)$," Proc. Amer. Math. Soc. (to appear).

21. I.G. MacDonald, "Duality over complete local rings," Topology I (1962) 213-235.

Offsetdruck: Julius Beltz, Weinheim/Bergstr.

Lecture Notes in Mathematics

Bisher erschienen/Already published

Vol. 1: J. Wermer, Seminar über Funktionen-Algebren. IV, 30 Seiten. 1964. DM 3,80

Vol. 2: A. Borel, Cohomologie des espaces localement compacts d'après J. Leray. IV, 93 pages. 1964. DM 9,–

Vol. 3: J. F. Adams, Stable Homotopy Theory. 2nd. revised edition. IV, 78 pages. 1966. DM 7,80

Vol. 4: M. Arkowitz and C. R. Curjel, Groups of Homotopy Classes. 2nd. revised edition. IV, 36 pages. 1967. DM 4,80

Vol. 5: J.-P. Serre, Cohomologie Galoisienne. Troisième édition. VIII, 214 pages. 1965. DM 18,–

Vol. 6: H. Hermes, Eine Termlogik mit Auswahloperator. IV, 42 Seiten. 1965. DM 5,80

Vol. 7: Ph. Tondeur, Introduction to Lie Groups and Transformation Groups. VIII, 176 pages. 1965. DM 13,50

Vol. 8: G. Fichera, Linear Elliptic Differential Systems and Eigenvalue Problems. IV, 176 pages. 1965. DM 13.50

Vol. 9: P. L. Ivănescu, Pseudo-Boolean Programming and Applications. IV, 50 pages. 1965. DM 4,80

Vol. 10: H. Lüneburg, Die Suzukigruppen und ihre Geometrien. VI, 111 Seiten. 1965. DM 8,–

Vol. 11: J.-P. Serre, Algèbre Locale. Multiplicités. Rédigé par P. Gabriel. Seconde édition. VIII, 192 pages. 1965. DM 12,–

Vol. 12: A. Dold, Halbexakte Homotopiefunktoren. II, 157 Seiten. 1966. DM 12,–

Vol. 13: E. Thomas, Seminar on Fiber Spaces. IV, 45 pages. 1966. DM 4,80

Vol. 14: H. Werner, Vorlesung über Approximationstheorie. IV, 184 Seiten und 12 Seiten Anhang. 1966. DM 14,–

Vol. 15: F. Oort, Commutative Group Schemes. VI, 133 pages. 1966. DM 9,80

Vol. 16: J. Pfanzagl and W. Pierlo, Compact Systems of Sets. IV, 48 pages. 1966. DM 5,80

Vol. 17: C. Müller, Spherical Harmonics. IV, 46 pages. 1966. DM 5,–

Vol. 18: H.-B. Brinkmann und D. Puppe, Kategorien und Funktoren. XII, 107 Seiten. 1966. DM 8,–

Vol. 19: G. Stolzenberg, Volumes, Limits and Extensions of Analytic Varieties. IV, 45 pages. 1966. DM 5,40

Vol. 20: R. Hartshorne, Residues and Duality. VIII, 423 pages. 1966. DM 20,–

Vol. 21: Seminar on Complex Multiplication. By A. Borel, S. Chowla, C. S. Herz, K. Iwasawa, J.-P. Serre. IV, 102 pages. 1966. DM 8,–

Vol. 22: H. Bauer, Harmonische Räume und ihre Potentialtheorie. IV, 175 Seiten. 1966. DM 14,–

Vol. 23: P. L. Ivănescu and S. Rudeanu, Pseudo-Boolean Methods for Bivalent Programming. 120 pages. 1966. DM 10,–

Vol. 24: J. Lambek, Completions of Categories. IV, 69 pages. 1966. DM 6,80

Vol. 25: R. Narasimhan, Introduction to the Theory of Analytic Spaces. IV, 143 pages. 1966. DM 10,–

Vol. 26: P.-A. Meyer, Processus de Markov. IV, 190 pages. 1967. DM 15,–

Vol. 27: H. P. Künzi und S.T. Tan, Lineare Optimierung großer Systeme. VI, 121 Seiten. 1966. DM 12,–

Vol. 28: P. E. Conner and E. E. Floyd, The Relation of Cobordism to K-Theories. VIII, 112 pages. 1966. DM 9.80

Vol. 29: K. Chandrasekharan, Einführung in die Analytische Zahlentheorie. VI, 199 Seiten. 1966. DM 16.80

Vol. 30: A. Frölicher and W. Bucher, Calculus in Vector Spaces without Norm. X, 146 pages. 1966. DM 12,–

Vol. 31: Symposium on Probability Methods in Analysis. Chairman: D.A.Kappos. IV, 329 pages. 1967. DM 20,–

Vol. 32: M. André, Méthode Simpliciale en Algèbre Homologique et Algèbre Commutative. IV, 122 pages. 1967. DM 12,–

Vol. 33: G. I. Targonski, Seminar on Functional Operators and Equations. IV, 112 pages. 1967. DM 10,–

Vol. 34: G. E. Bredon, Equivariant Cohomology Theories. VI, 64 pages. 1967. DM 6,80

Vol. 35: N. P. Bhatia and G. P. Szegö, Dynamical Systems: Stability Theory and Applications. VI, 416 pages. 1967. DM 24,–

Vol. 36: A. Borel, Topics in the Homology Theory of Fibre Bundles. VI, 95 pages. 1967. DM 9,–

Bitte wenden / Continued